W0081673

Study and Communication Skills for the Biosciences

Study and Communication Skills for the Biosciences

FOURTH EDITION

Stuart Johnson

Jon Scott

Dominic Henri

OXFORD
UNIVERSITY PRESS

OXFORD
UNIVERSITY PRESS

Great Clarendon Street, Oxford, OX2 6DP,
United Kingdom

Oxford University Press is a department of the University of Oxford.
It furthers the University's objective of excellence in research, scholarship,
and education by publishing worldwide. Oxford is a registered trade mark of
Oxford University Press in the UK and in certain other countries

© Stuart Johnson, Jon Scott, Dominic Henri 2025

The moral rights of the authors have been asserted

Third Edition 2019
Second Edition 2014
First Edition 2009

All rights reserved. No part of this publication may be reproduced, stored in a retrieval system,
transmitted, used for text and data mining, or used for training artificial intelligence, in any form or
by any means, without the prior permission in writing of Oxford University Press, or as expressly
permitted by law, by licence or under terms agreed with the appropriate reprographics rights
organization. Enquiries concerning reproduction outside the scope of the above should be sent
to the Rights Department, Oxford University Press, at the address above.

You must not circulate this work in any other form
and you must impose this same condition on any acquirer

Published in the United States of America by Oxford University Press
198 Madison Avenue, New York, NY 10016, United States of America

British Library Cataloguing in Publication Data
Data available

Library of Congress Control Number: 2024943609

ISBN 978–0–19–889136–9

Printed in the UK by
Bell & Bain Ltd., Glasgow

Links to third party websites are provided by Oxford in good faith and
for information only. Oxford disclaims any responsibility for the materials
contained in any third party website referenced in this work.

The manufacturer's authorised representative in the EU for product safety is
Oxford University Press España S.A. of el Parque Empresarial San Fernando
de Henares, Avenida de Castilla, 2 – 28830 Madrid (www.oup.es/en).

Preface

The transition from school or college to university is a very significant life event: for all students it marks the beginning of the next stage in their educational career and, for many, it may be their first experience of living away from home for any length of time. This transition, therefore, is associated with a whole range of new experiences in social and academic terms. Among these is a marked change in learning and teaching styles, with an expectation of increasing independence as a learner and development of a more mature, critical approach to the subject. Meeting the challenges posed by these changes requires the development of new skills in many areas of life, not least in the areas of study and communication. If you can develop effective study and communication skills early in your university career, you will facilitate your overall learning and help maximise your academic performance, as well as help to ensure you make the most of your degree.

This book was first developed from nearly 10 years of experience in developing and delivering a first-year module in Study and Communication Skills for the students in the School of Biological Sciences at the University of Leicester. Since the last edition was published, learning and teaching in higher education have changed significantly. For example, a much greater proportion is online, in digital formats and, of course, AI is proving to be a real game changer. When writing this, our fourth edition, we changed our approach with a focus on the student journey. In doing so, we thought about where students experience difficulty in developing the necessary skills and how best to support you to study effectively, approach assessments strategically, and be successful in your higher education career and beyond. We hope that future generations of students will be able to benefit from this guidance and we also look forward to learning more about the ways in which skills can best be developed.

S. J., J. S., and D. H.
Universities of Bristol, Leicester, and Hull.

Acknowledgements

We would like to thank all the Biology students who have helped to shape our work and, over the years, have experienced the evolution of our approaches to supporting their learning. Particular thanks are due to Katie Wray, our Editor at Oxford University Press for her extensive support and for keeping us on track throughout the production of this 4th edition. Thanks are also due to Jonathan Crowe who first conceived the idea of this book and has been a great supporter throughout.

Table of contents

Stage 1

Getting the most out of your course

Your initial thoughts may be: 'I did study skills at school so why do I need to read this book?'

Clearly, if you have got as far as getting a place at university you must be good at studying. Also, you were probably taught about study skills at school or college, and you will have certainly developed your own study skills during your time there. So, why do you need to read this book? The main reason is this: studying for a degree at university is different from studying at pre-university level. There are numerous ways in which it's different. The important ones are as follows:

- Studying at university requires you to be more responsible for your own progress than you were at school. You will, therefore, be expected to undertake much more self-directed study and so will need to learn to *manage your time well*.

- Studying at university requires you to be more critical about your subject than you were at school, and not just accept things at face value without question. You will be expected to identify differing views on a topic and develop your own views too. So, you will need to learn to *think critically*.

- Studying at university requires you to be more self-reflective about your ability and performance than you were at school so you can adapt and develop your current skills to better suit an undergraduate level of study. This means reflecting on your work and engaging with feedback to improve what you do in the future. So, you will need to learn to *develop yourself*.

In this first stage of the book, we will be thinking about the ways in which you can make sure that you get the most out of your course so that you work effectively and efficiently to achieve good results. In the first chapter, we will discuss the transition to undergraduate study and the significance of the foundational skills: managing your time, thinking critically, and developing yourself. We will then explore these aspects in more detail, starting off in Chapter 2 by looking at how you get to know about your course. Here, we will discuss the structure of Bioscience courses at university and how to engage with them. In Chapter 3, we will think about how to make the most of taught sessions and follow on in Chapter 4 with thinking about what to do when you are not in your taught sessions: the processes of independent study that are a key feature of succeeding at university. As we progress through this first section, we will also discuss how students with different approaches to learning can benefit most effectively from their studies.

1

The transition to higher education

Introduction

Studying at university is, for most, a more challenging experience than school. This chapter will focus on the differences you are likely to experience, and we will highlight what we consider to be the foundational skills you need to develop (and keep in mind) throughout your studies.

1.1 The contextual nature of skills: making the transition to undergraduate study

Skills are contextual. By this we mean that skills are learnt in a particular context, and the way you apply them is specific to that context. For example, when you learn to play a sport or to play a musical instrument you learn in a certain way. If you want to become proficient, you'll need some specific instruction, usually from a coach or a teacher, or by learning from peers who are already accomplished. You will be much more likely to become proficient if you're learning in an environment that is predictable and controlled, and so minimises some of the pressure of learning. For instance, you'll be encouraged to just 'have a go'; the emphasis will be on enjoying it rather than competing or performing. As you progress in your competence you may become very good at the sport or the musical instrument in the context of your learning environment (with people to help and a context that's familiar), but at some point, if you want to really improve, you will need to put yourself in more unfamiliar environments. If you really want to excel, if you're going to perform well on the pitch, the court, or the stage, you will need to step up and challenge yourself in new contexts.

Undergraduate study is a new and challenging context. You already have study and communication skills, probably ones that have served you well so far, but, as we have seen, studying for a degree at university is different from studying at pre-university level. Your skills will therefore need adapting and developing to fit the new situation you are encountering. The good news is that skills can be transferred from one context to another, so the skills that you have developed so far are by no means wasted. In fact, quite the opposite; they provide a very useful starting point.

1.2 Foundational skills

As we mentioned earlier, the important differences between studying at pre-university and at undergraduate level reveal three foundational skills that are vital for students who want to perform well during their degree: managing your time, thinking critically, and developing yourself. We unpack each of these skills (and many others too) in the chapters that follow, but these foundational skills underpin the ones that follow and so are worth drawing your attention to at this early stage.

1.2.1 Manage your time well

Many students, on nearing the end of their degree course, wish that they could have their time again. This is often because they feel that, given another go, they could do better. The reasons for thinking they could do better may be quite diverse, but common to many of them will be the notion that if they had managed their time more effectively, they would have been able to perform more effectively, and get a better degree. Therefore, managing your time well is crucial to performing well at university. There will be many demands on your time—social as well as academic—and you will have more autonomy about how you choose to spend your time compared with school or college. It's important, therefore, that you learn to allocate your time appropriately. This is one of those things that is much easier to say than to do; in theory it's straightforward, in practice it's difficult. Managing your time well is a foundational skill because it is necessary for performing well in a wide range of academic tasks (and, in fact, life generally).

1.2.2 Think critically

The second foundational skill that we want to draw your attention to in this opening chapter is the need to learn to think critically. Critical thinking is common to all academic endeavours and so it is important to understand what it is. Many people think of 'critical' as a negative term, and perhaps words like 'unfavourable', 'fault-finding', or even 'unkind' spring to mind. However, this isn't what is meant by being critical in an academic context. If you have ever read a review of a film, book, play, or concert, the review will have been written by a 'critic': someone whose professional job is to be critical in its full sense. For example, the critique of the performance of a play comments on all sorts of aspects of that play, including the quality of the performances of the individual actors, the way the stage was designed, the lighting, and so on. Thus, the critic may have written in glowing terms about the acting lead but felt that the supporting cast was not up to the job. In this context, being critical may mean being very positive and negative simultaneously about different aspects of the same thing. However, it is important to note that this kind of criticism is subjective: one critic's view of the play may be very different from another's.

In the academic sense, the word 'critical' also doesn't have specifically negative connotations. In fact, in an academic context, being critical is a good thing and is something to be encouraged. However, there is one key difference between thinking critically in scientific terms and in terms of reviewing a play: in science, the critical thinking should be undertaken objectively. For example, if you are comparing two theories, you should compare the ideas being presented and evaluate the weight of evidence supporting them, using this evaluation to decide which is most tenable. You should approach this evaluation from a neutral (impartial) viewpoint: not looking for something to be right or wrong because you think it should be, but because the available evidence indicates it to be so. Critical thinking is therefore a process that requires the evaluation of evidence to come to a conclusion that is supported by the outcome of experiments or observations. It is an intellectual exercise that is fundamental to the way in which science works and develops: the basis of scientific method.

We address how to apply and evidence critical thinking in more detail in Chapter 12, 'Key competencies: collaboration and critical thinking'.

1.2.3 Develop yourself

The third and final foundational skill that we have identified is the need to learn to develop yourself. This won't be an unusual concept to you as doubtless you will have been encouraged to engage in some kind of personal development planning at school or college. To make the transition from pre-university to undergraduate level, however, you will need to continue to develop yourself. Most universities will have a personal development planning scheme (or equivalent) in which you will be encouraged to take part. Such schemes are designed to help you in two ways: to improve your academic performance, and to help you make plans for what you will do once you have graduated. It is a process of continuous improvement: thinking about and reflecting on what you have done in the past, specifically thinking seriously about your progress and engaging actively with the feedback you have received from your tutors and your peers. From these activities you can learn and influence in a positive way what you do in the future (asking yourself questions like 'what worked well?', 'what didn't go as planned?', 'how can I make sure that the things that didn't work well work better next time?').

Regardless of the format of the system you use, the important thing is that you reflect on your progress (to identify where you are doing well and where you need to improve) and make plans for your future development. You will have numerous opportunities for reflection, including:

- feedback on coursework;
- feedback on exams;
- conversations with your lecturers;
- conversations with your personal tutor;
- conversations with friends;
- time by yourself thinking.

It is important that you use these opportunities to assess how you are doing and then (just as importantly) decide what you are going to do about it. If you have done well on something, then celebrate it in some way, but also make sure you are aware of why you got a good mark so you can take that approach again. Equally, when you identify areas for improvement, plan what you need to do to improve.

1.3 **Make yourself employable**

The final element to highlight in this chapter is the importance of making yourself employable. This isn't simply one skill but rather the combination of a whole range of skills that are dealt with throughout the book. Not only is making yourself employable about many skills rather than one skill, it is also about experience too, and the ability to articulate these skills and experiences in a clear and convincing way to potential employers. Given the breadth and importance of making yourself employable, this is something that we have written a specific chapter on (Chapter 14). Making yourself employable might not be something that you have at the forefront of your mind when you start at university—but you should! Gone are the days when a degree almost guaranteed you a graduate job. There are now far more graduates looking for jobs and the current economic climate means that some sectors are particularly challenging. This doesn't mean that a degree isn't worth it (on average, graduates still have a much higher employment rate and earn higher salaries than non-graduates) or that there aren't graduate jobs out there (some sectors are growing rather than shrinking). But it does mean that finding a graduate job is much more competitive than it used to be and that you will need to work harder, and start earlier, at making yourself stand out from the crowd.

 Chapter summary

Good study and communication skills are vital if you are going to do well at university and in future employment. Undergraduate study is different from pre-university study in a number of important ways. The skills you have already gained up to this point will be useful, but you will need to develop and adapt them for the new context you now find yourself in. Managing your time well, thinking critically, and being committed to self-development underpin the other skills that we will now explore in the subsequent chapters. These, together with appropriate experience and the ability to articulate your strengths to potential employers, are what will help you not just to succeed in your studies but to succeed in your working life beyond university.

2 Know your course

Introduction

For many students, university can be quite overwhelming. The way that a degree is structured, taught, and assessed can be very different compared to your past experiences of education. Especially because the onus is on you, the student, to make sense of everything and to ask for help when you think you need it. This confusion can lead to students struggling with their degree, not because they can't understand the biological content, but because they are not sure what they are supposed to be doing with it.

The purpose of this chapter is to provide some background on how degrees are structured to help you approach your degree as an informed and engaged student. By the end of the chapter, we want you to be able to answer a seemingly simple question for any activity you are undertaking as part of your degree, whether that be a lecture, practical, exam, or any other activity—'Why am I doing this and how does it contribute to my course?'. Once you know the answer to that question, you can ensure that you make the most of every activity and make good decisions about how you engage with your degree.

2.1 The structure of a degree

It's often difficult to understand things without a structure or framework, understanding how your degree is structured will help you make sense of the bigger picture and help you understand why you have been asked to do a particular activity and how it contributes to your degree.

2.1.1 Modules: the building-blocks of a degree

In the UK, a Bachelor's degree comprises 360 credits of study, most often this is split over three years. Normally, every student attempts 120 credits a year, with each credit representing approximately 10 hours of study, so each year requires 1,200 hours of study. Traditional bioscience degrees in the UK are taught in two 15-week trimesters (12 weeks of teaching and three weeks of exams), which means universities expect that students will need to spend 40 hours a week studying to complete their degree. This structure is not the only one, so you may wish to redo the workload calculations based on how your degree is structured (for example, if you are part-time, you are likely to be completing only 60 credits per year, which should come out at 20 hours a week).

Each year of 120 credits is usually broken up into modules, these can be 10, 20, 30, 40, and even 60 credits. Each module covers a different aspect of the subject, and the degree is completed by passing a sufficient number of relevant modules to get to the magic number:

Figure 2.1 Modules—the building blocks of degrees

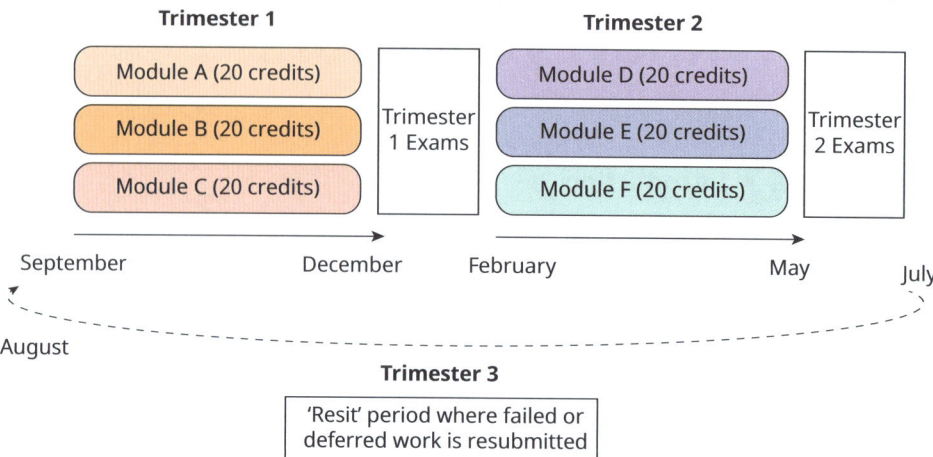

360. See Figure 2.1 as an example. Every bioscience degree at each university has a different structure, so it is not possible to say exactly how many modules you are taking and what topics they will cover; you will need to find out this information for yourself. This is because there is no centralised curriculum for most bioscience degrees. Universities are free to adjust their degree content, teaching methods, and assessment strategies to reflect the expertise of the staff who work there and the broader ethos of the degree (although degrees that are accredited by a professional body, such as the Royal Society of Biology or Institute of Biomedical Science, will have to teach certain topics and use certain assessment types in order to be accredited). However, there are some general rules that are likely to be true: some of your modules will be core and you will have to take them to complete the degree, and some will be optional modules that you can choose to take that allow you to tailor the degree to your interests. Usually, your first year of study will be mostly compulsory and your final year will contain more options.

Often, universities assume that their students will know this information already (which you may) but not everybody does, and it is important knowledge to help make sense of your studies. Below, we have provided an example of a common first year Bioscience structure and an attempt at visualising what all this information looks like in Figure 2.2.

Figure 2.2 Visualisation of module structure of the example year of study

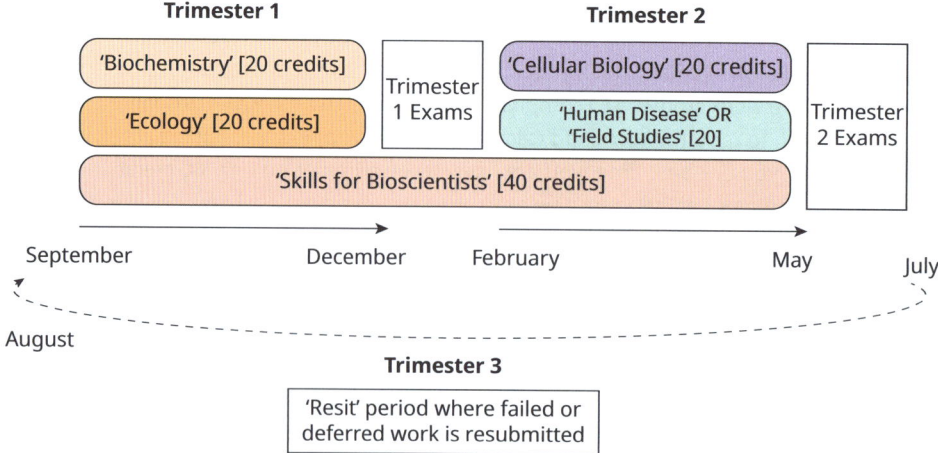

Table 2.1 Some key terms and their definitions (although note that these often vary between institutions)

Term	Definition	Common synonyms
Degree	The named Bachelor's qualification that is being studied for (e.g. BSc Biology). Consists of 360 credits of study from across multiple modules.	Programme
Module	A subset of the taught content of the degree. Contributes a set number of credits, which determines how much work is to be done and how much the final score on the module contributes to the final grade on the degree.	Unit
Core	A module that must be attempted and/or passed in order to earn a degree. Every student on the degree will take this module.	Compulsory
Optional	A module that can be chosen. Potential options are usually discussed during the academic year prior to taking the module. Represent an important opportunity to tailor your course to your aspirations and interests. Recommend discussing choices with your tutor.	Elective
Semester	A single term of study in an academic year that is broken into two blocks (semesters)	
Trimester	A single term of study in an academic year that is broken into three blocks (trimesters)	
Level	A grouping of 120 credits that most commonly refers to the year of study within the degree (e.g. 1st year, 2nd year, etc.). Note the numbering system is highly variable between universities.	Block
Lecturer	A member of academic staff who is responsible for some aspect of teaching on a degree. May have the honorific 'Dr.', which most often denotes that they have completed a PhD-level qualification.	Professor
Tutor	A member of academic staff who is supposed to be your primary point of contact for academic support during your studies.	Year lead, Academic Support Tutor

---

Basic 6 x 20 credit modules: First year BSc Biology degree [hypothetical]:

Trimester 1: Biochemistry (20 credits – core) AND Ecology (20 credits – core)

Trimester 2: Cellular Biology (20 credits – core) and EITHER Human Disease (20 credits – optional) OR Field Studies (20 credits – optional).

Trimesters 1 & 2: Skills for Bioscientists (40 credits – core)

Try this—know your course

Now when you are attending a taught session or completing an assignment, you should at least know how it fits into the big picture. Try creating a map of your modules for this year, which ones do you have to take, and which ones can you choose?

2.1.2 Assessment: the measure of engagement

Assessment performs a fundamental and central role in learning. When we are asking the question 'Why am I doing this and how does it contribute to my course?', we are fundamentally asking a question about how a learning experience is linked to the assessment. To engage with your degree effectively, you should know what you are going to be asked to do with the content of any given taught session. For example, what do you need to do with the knowledge covered during any given lecture, the data collected during a practical, or the discussion held during a seminar. This is not to say that there is no value in your course content beyond what it contributes to your final grade, there are plenty of rich life experiences and interesting subjects to be had during your studies. However, if we link back to our original point: 'students struggle with their degree, not because they can't understand the biological content, but because they are not sure what they are supposed to be doing with it', we think it is prudent that all students are aware that assessment is the fundamental measure of engagement with the degree. It is possible to attend every lecture and still fail if you don't submit the assessments.

Assessment is most often inherently tied to a specific module, and rarely to the degree as a whole. Each module you take will have a certain number of assignments that contribute to the final grade on that module. Typically, this will mean:

- Each assessment will contribute a set amount to one module, which is known as the assessment weighting (for example, a module could be assessed by two assessments that each contribute 50% to the final module grade);
- Some assessment is summative (it contributes to the final grade on the module) and some assessment is formative (it is designed to help you do better on later assignments but does not directly contribute to final module grades).
- The final grade on a module is equal to sum of (assignment score multiplied by the weighted contribution to module grade) for each assignment on that module.

2.1.3 University grading systems (What's a Tutu?)

For many students, receiving their first grade can be a big shock as 60% is the threshold for a 2:1 (pronounced 'two–one'), which is considered a 'good degree' by the UK government. A high-quality piece of work may obtain a grade of more than 70% (a First classification) and very few assignments will achieve a grade higher than 80% (still a First classification). The eponymous 2:2 (pronounced 'two–two') demarks a final grade of 50–59%. This is often a large jump down from the percentage-based systems you may have experienced before university.

The final grade for a degree is based on the weighted average grades across the summative levels of study. Often, but not always, the first year is formative (with a 40% pass mark), which means the grades remain on the transcript you might send potential employers but don't contribute to your final degree outcome. The second year contributes 20–40% of the

Table 2.2 Grading system for UK university Bachelor's degrees (note this is different from the grading system for Master's degrees)

BSc Undergraduate grading scale	
Final degree grade	Classification on degree certificate
70%+	First
60–69%	"2:1"
50–59%	"2:2"
40–49%	3rd
0–39%	Fail
MSc Postgraduate grading scale	
Grade boundary	Classification
70%+	Distinction
60–69%	Merit
50–59%	Pass

final grade for the degree (different for each institution). The final year will make up the remaining 60–80%. Again, the relative weighting of each year of study differs between universities, so make sure you know what the regulations are for your degree.

2.2 A strategic approach to engaging with your course

So, how should you engage with your course? We suggest a strategic approach, and it's helpful to consider this at the beginning of each trimester.

2.2.1 Developing SMART goals

There are many ways of approaching your degree, but a helpful method is to break it up on a trimester-by-trimester basis and tackle each trimester using project management techniques (see Chapter 5 for more detail). The first stage of effective project management is clearly identifying what it is you're trying to achieve.

So how do you know what you are trying to achieve? Try answering the following questions in order:

1. What modules am I taking this trimester?
2. For each module in turn:
   a. What are the assignments that I need to complete?
   b. How much does each assignment contribute to the final grade?

c. When do they need to be completed by?

d. How is the module taught?

e. How do the contents of the taught sessions relate to the assignments (e.g. are you collecting data in the laboratory session in a specific week that needs to be written up in a Lab Report)?

3. For students not in their first trimester only:

a. What feedback did I receive on similar assignments in the previous trimester of study?

b. Which modules from previous trimesters have content relevant to this trimester of study?

Below is an example of an assessment overview document for a single trimester of study (based on a hypothetical degree) that follows this method:

Module title: Biochemistry (20 credits – core):

Assignments details (including deadlines):

• Assignment 1. Multiple choice exam under exam conditions (50%)—date not available yet but during exam period in January.

• Assignment 2. Laboratory report (50%)—end of laboratory practical 1 in the final week of term (18 December). Note there is a practice (formative) version of the assignment due at the end of lab practical 2 in November.

Module teaching format and relationship with assessment: module content is taught via weekly two-hour lectures, the content of which is assessed in the multiple choice exam in January. No notes will be allowed, so start revision straight away (see Chapter 6 for guidance on exams and revision). There are also two laboratory practicals that each revolve around a problem-solving activity (see Chapter 3, section 3.3 for guidance on practical classes). These are assessed via the practice (formative) and final (summative) laboratory reports (see Chapter 8, section 8.3 for guidance on practical reports).

———————

Module title: Ecology (20 credits – core)

Assignments details (including deadlines):

• Assignment 1. Group presentation (25% – 5 mins)—Deadline in week 8 (20 November). There is a draft submission week 4 (23 October). Title—'Estimating population density of wading birds in a UK saltmarsh'. Need to find a group!

• Assignment 2. 1,500-word essay, titled '*Has ecology research contributed to better management of our environment?*'. (75%). Due final week of trimester (20 December).

Module teaching format and relationship with assessment: module content is taught via weekly two-hour lectures that have content relevant to the essay assessment. Not assessed via exam. What do I do during lectures? There is also a full-day fieldtrip. Need to present data collected in the presentation assignment and suggest how to use modern technology to improve the accuracy of future experiments (see Chapter 10, section 10.3 for guidance on scientific presentations).

———————

Module title: Skills for Bioscientists (40 credits – core, trimesters 1 and 2)

Assignments details (including deadlines):

- Assignment 1. Infographic (50%)—week 8 (21 November). I get to pick the topic! Note clash with Ecology presentation.

- Assignment 2. Data analysis activity (30%)—week 5 in semester 2 (no date yet).

- Assignment 3. CV and cover letter (20%)—final week of trimester 2 (no date yet).

Module teaching format and relationship with assessment: the module is taught in three batches of seminars focusing on digital communication skills, data analysis, and career management. Each of the batches contributes to one of the assignments, which are spread out across trimesters 1 and 2.

Once you have a clear statement of what you're trying to achieve (like the example above), you can create a specific set of goals. Project goals should always be SMART: Specific, Measurable, Achievable, Relevant, and Time-bound. If we carry on with the example above, you might notice that you have two assignments due in Week 8. It is your responsibility to make sure you are aware of deadline clashes and to manage your time to ensure you can put an appropriate amount of time into every assessment. It would be sensible to set some SMART targets to ensure that this happens as early as possible. Some example SMART targets for the hypothetical assessment overview document might look something like the following:

Ecology module presentation:

Week 1 by Friday—Read and analyse ecology presentation assignment brief (see Chapter 7)

Week 2 by Friday—Read and make notes on five papers that will be referenced in the assessment (see Chapter 9)

Week 3 by Friday—Finalise structure for all sections (see Chapters 8 and 10)

Week 4 by 23 October—Draft submitted to lecturer

Week 7 by Friday—Review feedback (see Chapter 13)

Week 8 by 20 November—Review and revise essay (see Chapter 11) and submit on VLE.

Skills for Bioscientists infographic:

Week 5 by Friday—Read and analyse ecology essay assignment brief, decide on topic

Week 6 by Friday—Read and make notes on at least five papers AND finalise layout for assessment (see Chapter 10)

Week 7 by Friday—Complete first draft

Week 8 by 21 November—Review and revise draft and submit to VLE.

If this was the assessment on your degree, then you would know that Weeks 7 & 8 are likely to be especially busy because there are at least two assignments that you need to work on during those weeks. However, you would have given yourself a much better chance of managing this workload because you made a start on the work from Week 1. It would be worth having a series of SMART goals for every assessment so you can plan your time wisely. Chapter 5 covers this kind of planning process in a lot more detail, but we strongly recommend that you make sure you know what assessment you have as early as possible in each trimester of study.

2.2.2 Where do I find this information?

Most of this information should be available on the Virtual Learning Environment within the first week of the start of the trimester (see Box 2.1 for guidance on VLEs). Often a lot of this information is covered on the first taught session of a given module as part of an introductory lecture for each module. Finally, there may be a Module or Student Handbook, or Assignment Briefing Booklet, that contains the required info somewhere on the VLE. If you cannot find what you need after exhausting all these avenues, it's time to ask for help! Try sending an email to the lecturer leading the programme or module, or even your tutor, asking where this information can be found. It might even be quicker and easier to stick around at the end of the next taught session on the relevant module to ask the lecturer leading that session where this information is usually stored?

BOX 2.1 Virtual Learning Environments (VLEs)

VLEs are widely used by universities to provide online information and academic content to support your learning. There are several commercial platforms used by universities, such as Blackboard, Moodle, or Canvas and some universities create their own bespoke systems.

Irrespective of the construction of the VLE used by your university, they are usually employed in similar ways to provide a guide to the structure of the programme and its constituent modules, as a system for managing the academic content and also as a means of communication. The VLE is also likely to host information about accessing academic and pastoral resources, for example the library and study-support functions of the university.

At the level of the individual module, you are likely to find specific information such as:

- the aims and learning outcomes;
- the timetable for the module;
- details of the assessments and links for submitting your assignments;
- a summary of the module content;
- links to resources such as reference papers and useful websites;
- recordings of any lectures or other materials made available online by your lecturers;
- communication regarding any updates to the module.

2.2.3 Why am I doing this?

Well done, you have completed the first stage of project management—knowing exactly what the project is! Now, whenever you are doing something as part of the degree, you should be able to answer a fundamental question that many students struggle with: 'Why am I doing this?'. You will find that the guidance in this book about 'how' to do something will make much more sense when you understand 'why' you are doing it. Let's consider an example of when understanding the 'why' makes for a more effective 'how' using the example course structure we have used throughout this chapter.

A student attending lectures for the Biochemistry module knows that the contents of those lectures are going to be assessed via an end-of-trimester, closed book exam (see Chapter 6, section 6.3.2 for explanations of different exam types). This means that they should attend the lectures with an approach that maximises understanding and recall of this information from memory, under timed and stressful conditions.

However, there is no exam for the Ecology module, instead the lecture content needs to be integrated into the 1,500-word essay titled 'Has ecology research contributed to better management of our environment?' This means that there is no need to memorise lecture content, in fact, that would be a waste of time. Instead, the student in question should be attending lectures with their essay plan to hand so that they can make note of which points made during taught sessions could and/or should be included in their essay. This means that they can focus on listening to, engaging with, and enjoying the lectures, with a greatly reduced focus on making detailed notes to aid future revision. Although, remember that some note-making will help with maintaining focus and understanding the content (which the student may well need to remember on future modules).

While both the Biochemistry and Ecology modules are taught in similar ways in this example, the student's approach to making the most of the taught sessions should be very different and based on an understanding of 'why' they are attending the lectures.

We recommend that you use this book as a guide to adopting a project management approach to your course, beginning each trimester by investigating what your goals are and then referring to each chapter in the book when you require help with the specific topic it covers (e.g. getting the most out of taught sessions). To help with this you could add to the list you have created above with the chapter and page numbers of sections of the book that provide guidance on the relevant type of taught session or assessment.

 ## Chapter summary

Hopefully, you should now have a clearer idea of how your course is structured. This knowledge should help you to engage with your degree from an informed position, make the most of your taught sessions, do well on your assignments, and help you make good decisions about how to use your time. You should also now be able to avoid one of the key barriers that impedes many students who struggle on their degree—knowing what you are expected to do! Having clear goals is a prerequisite to strategically engaging with your degree, which means you should be able to better balance your commitments outside of it; such as earning money, engaging with societies, family responsibilities, and building a healthy social circle. The rest of the book deals with advice on specific experiences we expect you to encounter as part of your degree. We recommend that you come back to this chapter at the beginning of each trimester to create your list of SMART goals.

3 Making the most out of taught sessions

Introduction

During your time at university, you will experience different formats of teaching. The approaches adopted will be selected to assist you in your learning and understanding of the subject. It is therefore important that you engage actively in all these different situations so that you can maximise the benefit that you gain from them.

The most common forms of teaching you are likely to encounter are:

- lectures (either in person or online as a live or recorded presentation);
- tutorials;
- practical classes.

In this chapter, we will consider each of these formats, thinking about how best to learn from them. Before we look at the details of the different types of teaching session, there is one common principle if you are going to make the most of them: preparation is key. If you go into a session not having prepared properly you will find it much more difficult to engage with the material and learn from what is being covered.

3.1 Lectures

Lectures are still a common feature of almost all undergraduate Bioscience courses; therefore, being able to make the most of them is an important skill. The lectures you have as part of your course will be many and varied; your lecturers will have different styles and different levels of ability in lecturing, and the lecture content itself will be of more or less interest to you, and more or less demanding. Also, the lectures may be delivered in person or as recordings posted on your Virtual Learning Environment (VLE). Regardless of these variables, however, you need to make the most of lectures.

Lectures can be a very valuable resource as they can synthesise the views of several researchers and textbooks, or provide new, even unpublished, information. Lectures are often liked by academics as a means of communication because they are a very efficient means (in theory at least) of transferring lots of information to large numbers of students all at the same time. How efficient a lecture actually is at communicating information will depend on at least three variables:

- how well the lecturer prepares and delivers the lecture;
- how well the lecturer engages the students throughout the lecture;
- how well the students are listening, making notes, and participating.

Clearly, you can't do much about the first two of these variables. The third variable (listening, making notes, and participating) is an area you *can* do something about; in fact, no one else can do it for you.

3.1.1 More than information transfer

Lectures are not merely about transferring information from the lecturer to the audience: if the purpose of a lecture was only to transfer information from the lecturer's brain to the students' notes, then lectures really have very little point, because there are many more efficient ways of doing this, such as providing handouts or suggesting particular chapters of a core text to read.

Lectures are of value because they add something in the communication of information that couldn't be achieved just as well by other means. You will know this if you have ever tried to read a handout or a friend's notes from a lecture that you didn't attend; trying to understand the information out of context can often be very difficult. The transfer of information is important, but it is what you do with the information as you receive it, and afterwards, that really matters. This is where you, as a student, come in. It is perfectly possible to disengage your brain, even when sitting in a lecture that has been well prepared and is being well delivered.

Doing something useful with the information as you receive it depends on processing the information effectively. This processing can't all take place within the lecture itself (even if you do listen actively and make notes appropriately), which is why this chapter covers preparation before lectures and follow up after lectures, as well as what to do during lectures. Before we address these issues, it will be helpful to consider how lectures are different from what you may have experienced at school or college, prior to your undergraduate degree.

3.1.2 Differences between lectures and lessons

Lectures at undergraduate level will be different from lessons you had at school or college in a number of important ways. At the risk of generalising, these differences probably include the following:

Class size

In years 12 and 13 at school, you may well have been in a class of fewer than 30 students; at undergraduate level it's not unusual, especially in the first year, for class sizes in face-to-face sessions to be up to 200–300 students (although typically, as you move through your course to the final year, the class sizes will get significantly smaller).

Anonymity

A bigger class makes individuals within the class more anonymous; it is easier in a larger class just to sit quietly and not really engage with the lecture content because you think no one will notice. This is also particularly the case if you are viewing the lecture online.

Relationship

Bigger class sizes or online learning mean that you are more remote from the lecturer, not only physically, but also in terms of relationship. The chances are that, in your first year of study, you probably won't get to know most of your lecturers very well, and they probably won't even know your name.

Completeness

In school, your lessons probably covered everything you needed about your syllabus. Importantly, at undergraduate level, lectures are not intended to tell you everything you need to know and so need to be supplemented with additional reading (but more on this one later).

3.1.3 Prepare before lectures

You may already be on a busy course with a lot of timetabled teaching hours so is preparing for lectures really necessary? We think it is. Preparing for lectures doesn't have to take long and can actually save you a lot of time. It doesn't have to take long because you are not trying to cover all the material the lecturer is going to cover, after all that's what the lecture is for. Instead, you are just giving yourself a framework to help you understand it better. It can save you a lot of time, because if you have a framework to help you understand the lecture, you can process the information more quickly, thus making the listening easier, the note-making more selective, and the follow-up more focused. Preparing for lectures makes for a much more productive learning experience. To know how to prepare for lectures, though, you need to know your course (see Chapter 2).

Reading before the lecture

We have already noted that the pre-reading is aimed at giving yourself a framework to help you understand the lecture better. This is because it gives you somewhere to place information and so relate it to other information, rather than trying to understand it in isolation (which is always very difficult).

The framework can be quick to create because you only need to understand basic structure and terminology, rather than detail. The lecturer may upload all the slides or lecture content onto the VLE beforehand; if so, have a look at them and download them so you can annotate them during the lecture. A good way of supplementing this is by looking up the relevant chapter in the core text for the module (this is where you need to know the title of each lecture) and scanning through it to familiarise yourself with the main themes. You don't need to read it in detail, just scan through all or some of the following:

- the chapter overview (if there is one);
- the chapter headings and subheadings;
- the introduction to the chapter;
- the conclusion or summary to the chapter;
- figures, diagrams, or tables (which often summarise a lot of information succinctly).

Also, look out for any terminology that you are unfamiliar with and try to find out what it means.

Core texts are not the only source of information for pre-reading, however; your own notes will be useful, too. Assuming you are not on the first lecture of a module, your notes from previous lectures in the module will be an important source of information. Again, you don't have to read all the notes in detail, just scan through them to remind yourself of what was covered previously. If you come to a section that you don't understand, spend more time on it and check up on the topic in a textbook or other reference source, or ask your friends. If you have prepared for a lecture, it makes listening and note-making much easier.

Pre-lecture activities

Your lecturers will want you to arrive at their lecture prepared for what they are about to deliver. Some of them may use formal pre-lecture activities to help you prepare. These may take many forms: they may ask you to read a section from the recommended textbook or to read ahead in the lecture notes; some lecturers will adopt a 'flipped' approach, where the lecture itself is made available as a recording beforehand and then they will use the lecture time to focus on some specific points or generate a discussion (see Box 3.1). If you haven't gone through the pre-lecture activities, you will quickly find that you are at a significant disadvantage. Some lecturers may start their lecture with a quiz or a few short questions, based on the previous session. All these activities are very useful to engage with and have been designed explicitly to help you understand what is coming in the lecture. So, take them seriously and always do any pre-lecture activities that are prepared for you.

BOX 3.1 The flipped lecture

Many lecturers now audio- or video-record their lectures. This has led to some different approaches to the traditional lecture. One of these is commonly referred to as 'flipped lectures'. As the name suggests the approach is very different to a traditional lecture where the lecturer gives information to the students in the lecture. In a flipped lecture this situation is reversed—you will be expected to have studied the content of the lecture in your own time, often by watching a video lecture or by studying complete handouts. The 'lecture' sessions are then used for students to be engaged in activities such as answering questions, perhaps using audience response systems, or discussing and sharing their knowledge and understanding with other students.

Make sure that you engage seriously with any pre-work that has been set. Do this actively and make notes and annotations for yourself (see section 3.1.5 *Make notes appropriately during lectures*). Then, when you arrive in the classroom or lecture hall, make sure that you attempt all the activities that are given to you. This is where your understanding will be consolidated and where you will have the chance to practise the types of activities that may be required for assessment. Clearly, this is a very different approach to the traditional lecture and many of the lectures that you have will actually fall somewhere between these two extremes, with a mixture of pre-lecture preparation and active engagement during the classroom sessions.

Try this—Preparing for a lecture

Read through the notes you took from the previous lecture and make sure that you understand all the material. If there are any aspects that you don't understand, check them in a textbook or ask your lecturer. Go through any pre-reading and make notes from it. If there isn't any pre-reading, skim-read the topic in a textbook so that you get an overview of what is to be covered.

When you have done this and have attended the lecture, think back on how well you felt able to follow the lecture compared with one where you have gone in unprepared. Try to get into the habit of preparing for every lecture.

3.1.4 Listen actively during lectures

This section is deliberately titled 'Listen actively' because there is a difference between just listening and listening actively. Listening actively suggests that you are alert, attentive, and ready to engage with the content of the lecture, as opposed to just being there. Listening actively also means that you are willing to work hard at listening, regardless of the quality of the lecture. Additionally, we have already identified that preparing for a lecture makes listening actively much easier, because you have a framework within which to place new information and you will at least recognise the terminology used.

During the lecture you are likely to be using electronic devices to take notes and will have other devices as well, so do make sure that you have switched off any other notifications, such as social media, that are likely to distract you.

Even when you have the best intentions to remain focused throughout the lecture, you may well find your concentration slips at times. The active engagement points we set out in the following paragraphs will help you to remain focused but if you do find you have missed a point, make sure you follow up after the lecture: see section 3.1.6.

Identify your priority

There will be some lectures when it will be difficult both to listen and to make notes, either because of the complexity or newness of information being communicated, or simply because of the amount. In such situations you will need to identify what your priority is—is it to listen or is it to make notes? Deciding which your priority is will depend on whether you are more concerned with trying to understand the information or collect information. If your priority is *understanding* the information then you will need to:

- focus on listening;
- make only brief keyword notes;
- if you have downloaded the slides your notes can be very brief annotations;
- follow-up the lecture by making detailed notes.

If, however, your priority is *collecting* information, then you will need to:

- focus on making notes and making detailed annotations on the slides;
- make more detailed notes;
- follow up the lecture by reviewing your understanding of the content.

When making a decision about your priority, also consider what resources are available to you after the lecture. It is usually relatively easy to get hold of appropriate textbooks and you may have access to lecture handouts or, more commonly, an electronic copy of the slides. Many universities will provide you with lecture capture facilities. Such facilities can be very helpful if used strategically, to review content, for example, you can go back over a section of the lecture where you had difficulty in keeping up or did not understand a specific point. They can also be very useful for revision. Importantly, however, they are not an excuse for non-attendance. Even if you can listen to a recording, it is rarely as good as actually being there. This would suggest that active listening and, therefore, understanding should be more important than making notes. Remember, however, you do not necessarily have to keep to the same strategy all the way through a given lecture; you could switch between the two depending on the material (more on this in section 3.1.5 *Make notes appropriately during lectures*).

Listen for structure

Following a lecture and making notes are always easier if you have an awareness of the structure of the lecture. Being aware of structure enables you to be more selective and, therefore, more focused, in your note-making because it gives you an indication of what the important bits are. That is not to say that the other bits are unimportant, it's just the recognition that, in terms of understanding, there are certain elements of a lecture that are helpful to grasp in order to understand the rest of it.

Sometimes the structure of a lecture is very clear, other times less so. It is important, therefore, to know what the structural clues might be. The most obvious, and probably the most common way of making the structure of a lecture clear, is for the lecturer to outline what the structure is going to be at the beginning of the lecture, for example as a list of the learning outcomes or topics. If a lecturer does outline the structure, make sure that you make a note of it straight away, as it will give you a sense of direction and help you to anticipate points or take up the thread of information again, should you get lost.

Additionally, during a lecture, the lecturer will probably give you some cues or 'verbal signposts', these include statements such as:

- 'I shall now discuss ...'
- 'My next point is ...'
- 'Finally ...'
- 'In conclusion ...'
- 'To summarise ...'

These signposts identify a new point or stage in the lecture, and you should show this in your notes accordingly. Other signposts include pausing to indicate a new point or summarising what has been said prior to moving on.

There are other, more subtle, verbal signposts, which can help you structure your notes; you will need to listen for these. Examples include:

- 'On the other hand ...'
- 'Others have argued ...'

- 'Turning now to …'
- 'Alternatively …'

Other words and phrases indicate that an illustration is about to be given:

- 'An example of this is …'
- 'This can be seen when …'
- 'Evidence for this can be found in …'

Your ability to listen actively and, in particular, to listen for structure, will improve with experience. As you improve, you will be better able to spot digressions or additional examples, and adjust your note-making accordingly, which brings us on to the next section.

Engage with in-lecture activities

Your concentration levels are very likely to drop off at times during a lecture. So, many of your lecturers will build activities or breaks into their lectures to help you refocus and regain your concentration. These lecture activities can take a variety of forms, depending on the nature of the topic being taught, the constraints of the lecture room, the size of the class, and the preferences of the lecturer. Whatever these activities are, you will find them more useful if you engage with them actively.

During an in-lecture activity you might be asked to:

- discuss something with your neighbour and feed back;
- stand up if you …;
- raise a hand if you …;
- vote using coloured card or raising your hand;
- watch a video;
- write a question on a post-it note;
- answer multiple choice questions using an audience response system.

If the lecturer asks a direct question to the class as a whole, it can be scary to suggest an answer but do try: it is natural to think that other people will be more confident and will know more than you but in reality that is unlikely to be the case, particularly if you have done your preparation beforehand. Likewise, if there is a point that you don't understand you can raise your hand and ask: if you don't understand it, it is a certainty that others of your classmates won't have understood it either.

If you find it hard engaging with questions in large classes, it is always worth talking with the study skills advisers about approaches that can be helpful in overcoming these barriers.

3.1.5 Make notes appropriately during lectures

We use the phrase 'make notes' rather than 'take notes' because the process of making implies an active, creative process, rather than passive taking. In the case of lectures, it is important that you create notes that are unique to you and are produced by you engaging thoughtfully with the content of the lecture.

Know why you are making notes

We have already encouraged you to think about what your priority might be when in lectures, and understanding information and collecting information are two key reasons for making notes. However, there are other reasons, too. Making notes appropriately in lectures can also help you to:

- concentrate better;
- remember the content better (in the short term at least);
- think about questions you may want to ask or highlight points you don't understand;
- highlight areas of interest.

Know how to make good notes

The ability to make good notes is a skill that develops with practice, so don't expect to be an expert at it straight away. Clearly, you will have made notes before at school or college, but making notes in lectures in an undergraduate setting is different for a number of reasons, as highlighted in section 3.1.2 *Differences between lectures and lessons*. Note-making is also a very useful skill to develop for other forms of study, whether that be teaching sessions such as tutorials or practice classes, as we will discuss later, or note-making from your reading of texts or research papers. Improving your note-making skills comes partly through practice, but there are some important principles to bear in mind, too. These include:

- using structure in your notes;
- using your own words;
- using fewer words;
- using abbreviations;
- using space;
- using colour and image;
- using handouts or downloads of the slides;
- organising your notes.

Use structure in your notes

During the lecture you may write your notes by hand or type them but notes that are lacking in structure will be much more difficult to understand, especially after the lecture, than notes that have a good structure. Imagine you were looking at some lecture notes a few days after a lecture that looked like the ones in Figure 3.1. Notes like these are difficult to understand, partly because they are lacking in structure.

Alternatively, the notes represented in Figure 3.2 have better structure and so should be much easier to understand. The main purpose of structure is to make clear which information is important. It is crucial to note that identifying important information is much easier during the lecture because lecturers will use their tone of voice, pace, and many other devices to provide emphasis to the material, thus giving you an indication of what aspects of

Figure 3.1 Poorly structured lecture notes

Respiration - Prof Smith   22 Feb

Inspiration - depends on contraction of diaphragm - flattens - and of external intercostals. Cause increase in thoracic vol., reduces pressure, draws air in.
Expiration - passive process - elastic recoil of lung tissue. Air flow is tidal, air in lung is not the same composition as air outside - has more CO_2, less O_2 'cos some remains in lung at end of expiration - even if forced. Inspiration has to do work - overcome elastic tension of lung tissue and surface tension of fluid lining lungs. ?role of surfactant??
Gas exchange - occurs in terminal bronchioles and alveoli. O_2 dissolves in fluid and diffuses across lung wall to blood vessels - down concentration gradient. O_2, poorly soluble so needs large surface area for dissolving. CO_2 is better, goes in opposite direction. Need short diffusion path.

Figure 3.2 Well-structured lecture notes

Respiration - Prof Smith   22 Feb

Inspiration*
┌─ Dia. + ext intercostals contract
├─→ thoracic vol ≠ + pressure ø
└─→ air drawn in
* has to do work - overcome elastic tension of lung tissue and surface tension of fluid lining lungs. ?role of surfactant??

Expiration
- passive process - elastic recoil of lung tissue
- air flow is tidal
- air in lung is not the same composition as air outside - ≠ CO_2, ø O_2 (some remains in lung at end of exp.)

Gas exchange
- occurs in terminal bronchioles + alveoli
- O_2 dissolves in fluid and diffuses across lung wall to blood vessel - down concentration gradient
- O_2, poorly soluble so needs large surface area for dissolving
- CO_2 is better, goes in opposite direction. Need short diffusion path.

the information are most important. After the lecture, however, it is very difficult to recall this kind of detail. To give your notes structure, you should:

- use headings to order information;
- give each point a new line;
- highlight examples and illustrations in an appropriate fashion;
- use diagrams to summarise information;
- make clear when sections of your notes are digressions from the main points.

Use your own words

One of the reasons why it is important to try to use your own words, when making notes in lectures, is because it will help you (or perhaps force you) to understand the content of the lecture better. If you are trying to put information in your own words, then you must process it as you hear it in order to express it in your own words. You may have experienced occasions in lectures when you have been able to write down what the lecturer is saying without actually thinking about what is being said—this illustrates how passive the process of note-taking can actually be. How much you try to put information into your own words will depend, to a certain extent, on your priority, but whether your priority is to understand or collect information, putting that information in your own words will help you understand it better. This does not mean that you have to put absolutely everything in your own words; for example, there are two particular occasions when it is important to record the precise wording:

1. *When you are recording a quotation that the lecturer is referring to*: in this case make this clear in your notes by using quotation marks.
2. *When you don't understand what the lecturer's words mean*: in this case make this clear in your notes by adding a question mark in the margin (for example) as a reminder to follow up the point later.

Know how much to write down

There are two potential problems related to how much information you write down—writing down too much information or not writing down enough. How much you write down will, again, depend on your priority (understanding information or collecting information), but it is also influenced by your own level of confidence. Under-confidence tends to lead to writing down too much, whereas over-confidence tends to lead to not writing down enough; neither is ideal. Among first-year undergraduates, probably the more common tendency is to write down too much information, but whichever your tendency, the following suggestions will help:

- remember that a lot of notes doesn't necessarily equal good notes;
- look and listen for the important points—these are often the structural parts;
- use keywords to represent points or ideas concisely;
- add brief details of any examples or evidence that support a point.

Use abbreviations

Using abbreviations can be a real time-saver. Use standard abbreviations, subject-specific abbreviations, and your own abbreviations for common words. The important thing is to be consistent and to ensure that your notes are still comprehensible. Don't use so many abbreviations that your notes turn into a shorthand transcript—these can be very difficult to decipher when your memory of the lecture has faded.

Use space

It can be tempting to try to cram as much information onto a page or screen as possible, but this will create difficulties for you both during and after the lecture. Notes that are densely packed are difficult to review and difficult to make additions to at a later stage. It is, therefore, helpful to use space in your lecture notes to make them easier to review and easier to supplement with additional material. To create space, make sure you put each point on a new line (this also helps represent structure) and leave gaps for additions or corrections, especially if you think you may have missed a point or don't understand something.

Use colour and diagrams

We have already identified the value of highlighting important points by using structure in the section on '*Use structure in your notes*', but this can be further enhanced by using colour and diagrams, too. It can be useful to highlight key points in colour, and a quick sketch as a way of summarising a concept or idea. Sometimes such images or diagrams will be used by the lecturer, in which case, if it is useful, copy it down, although you are very likely to be provided with the slides to download, so you do not need to be able to draw well (see also the points in '*Use slides and handouts effectively*'). On other occasions, you will think of ways to visually represent something that is only communicated verbally; this can also be a very helpful thing to do, but make sure you record sufficient information to be able to understand the concept or idea at a later stage.

Use slides and handouts effectively

Slides and handouts are a very valuable source of information. Most lectures won't have handouts, but many will have slides. Assuming that copies of the slides are available to you (either before or after the lecture) then the principles of using them effectively are similar.

- When making notes, think about what is already provided, and don't copy down in your notes things that are already there. This will enable you to listen more actively and so focus on understanding.

- Make notes on the handouts or slides themselves, whether they are paper or digital— highlight important points, annotate, add comments, and write down any questions you have.

- Some lecturers will provide 'gapped' handouts or slides where you need to fill in the gaps with key information. Follow the lecture so that you fill in the correct information in the gaps.

- Don't fall into the trap of thinking that because you've got a handout or slides that you don't need to listen much—listening actively during the lecture and making appropriate notes will save you a lot of time later and also make it easier to understand the material.

- File your handouts and notes in an organised fashion (which brings us neatly onto the penultimate point in this section).

Organise your notes

It is important that you organise your notes effectively (whether paper or digital), as this will make it much easier for you when you follow up the lecture at a later stage, or use the notes for background reading for an essay or when revising for exams (see Chapter 5 *Getting yourself organised*). You may well be making your notes digitally on a laptop or tablet. If you do this, then the same principles regarding structure, your own words, abbreviations, and so on still apply. Using a laptop or tablet also has the advantage that notes can be tagged, retrieved, and searched in a way that you can't do with paper-based notes (as examples, you can check out Evernote, Notability, Google Docs, Microsoft OneNote, and GoodReader). When using a laptop or tablet to make notes, you may need to work harder initially to ensure you still use space, structure, colour, and image (to avoid your notes being linear and characterless), but this will improve with practice.

If you are making notes on paper, the simplest method is to use A4 paper for taking notes and to store your notes in a ring binder, together with the relevant handouts. To make ordering your notes easier, begin notes for each lecture on a new piece of paper and give them a clear heading of the lecture title, date, and name of the lecturer. Also add page numbers so you can order the pages easily. To store your notes and handouts in a ring binder you might need to hole-punch them—do this before they get lost.

Try this—Comparing lecture notes

After a lecture, compare your notes with other people. Remember, we said earlier that the ability to make good notes is a skill that develops with practice; it is also a skill that develops as you see how other people do it. Comparing notes can be a helpful exercise because it can help you to:

- see how other people take notes and pick up good techniques;
- identify and fill in any gaps that you might have in your notes;
- discuss the content of the lecture and clarify your understanding;
- identify how the lecture relates to the rest of the module.

Know the common problems and how to address them

We have addressed the ways to help you to make appropriate notes, but there will always be times when, for a variety of reasons, concentration dips and you begin to get left behind in the lecture. The trick here is not to despair and give up.

Failing concentration

If you are using a laptop/tablet/phone during the lecture, make sure you are not tempted by distractions by switching off social media and emails, etc. You are much less likely to find your concentration straying when you use an active approach to note-making. By using your own words, and by using space, colour, and image, note-making will become a busy but interesting activity. If you do miss some points because your attention strays, then just leave a space in your notes and check it out with the lecturer or another student later or, if lecture capture is available, review the recording of the lecture.

Being left behind

In a live lecture you may find that information is being delivered too fast for you to write down. If points pass you by, then again, leave a space and compare your notes with another student's or check the recording. Remember that doing some background reading for the lecture will help you to keep up as the information will not be entirely unfamiliar to you. Sometimes you can get lost because you don't understand the material that is being delivered. This may be the case for the occasional point or even for a large section of the lecture. Rather than giving up on the lecture, write a series of questions that you can try to follow up later.

3.1.6 Follow up after lectures

After the lecture, it is tempting just to file everything away and move on, but we would encourage you to take some time to follow up on the lecture to ensure that you have understood everything, that your notes make sense and that you can place the lecture in the context of the topic you are studying. The best ways to do this are to:

- re-read your notes and ensure you understand them, highlight the important points—and any points that you don't understand as a prompt to follow up;

- if there are recordings made available after the lecture, do use these to recap any sections that feel you may have missed or don't understand. One of the great advantages of the recordings is that you can skip through to specific points rather than sit through the whole lecture again (you can also increase or decrease the playback speed);

- most lecturers will provide reading lists that place the lecture content in the context of the topic and allow you to explore the subject in more detail. The extent to which you do this is up to you but do engage with this reading as it will help you understand the topic now and notes you make from the reading will also be a very useful resource when it comes to revision;

- don't be afraid to ask questions: talk the topic through with your friends or make contact the lecturer with a specific question.

3.1.7 Online lectures

Many programmes are delivered through a mix of in-person and online sessions. Lecturers may pre-record lecture content for you to view online in your own time. In-person sessions are often also recorded and made available after the event. These recordings can be very useful in allowing you to go back over a lecture to review something you may have missed or not understood. But don't use the availability of recorded lectures as an excuse for not attending a lecture in-person.

If you are following a lecture online, make sure that you are in a quiet environment where you won't be distracted by other things going on around you; do also do make sure that you have turned off your social media apps, emails, and other websites as it is very easy to be distracted by messages or posts and engage with them rather than the lecture. As far as possible, you should aim to focus your attention just as much as if you were in the lecture theatre and engage with the material, for example, by taking notes and identifying any points that you don't understand so you can follow up after the session.

3.2 Tutorials

Tutorials form key elements of teaching within most degree programmes and can be employed in a variety of formats, as we will see in this chapter. Irrespective of the specific format of a session, the most important things you can do are to *prepare* and to *participate*. Without preparation and if you are not willing to participate, you will derive little benefit from a tutorial.

Tutorials can be quite a daunting prospect, as many students are unsure of their knowledge base and ability to discuss ideas, particularly in front of a member of staff who may be a world expert on the subject. But you should remember that your fellow students will be in a similar position and your tutor will want to encourage you to participate. If you find that participating in tutorials is too daunting, then try arranging to speak with your university's study skills advisers who can help you overcome some of these challenges.

Tutorials are useful in developing a range of important skills that will be of value to you in your studies and also in your career development. These skills include the development of:

- *subject awareness*—increasing your knowledge of the subject;
- *critical thinking*—listening to different ideas as they are put forward, thinking about them, and evaluating them in terms of the strengths and weaknesses of the arguments;
- *communication skills*—presenting your own ideas in a way that can be understood by the rest of the group, and learning the essential ability to refine the views expressed by other people, pick out the key points, or even counter them, without causing offence.

These skills, particularly the critical thinking and the communication skills, link closely with the skills that are sought by graduate employers (see Chapter 14 *Making yourself employable*), so developing them is beneficial both for your immediate academic performance and also for your future career prospects.

3.2.1 Preparing for academic tutorials

Preparation for tutorials is essential. It may well be that you have been given a specific topic to research, a worksheet to go through, or some sample questions to answer. On the other hand, you may simply have been told what the theme of the tutorial will be. Either way, you need to be prepared, so that you can benefit from the discussions and participate in a meaningful way.

Tutorials offer you the opportunity of discussing your subject with a bioscientist, who is very knowledgeable, as well as with your fellow students, in a fairly informal setting. If you are going to be able to engage in that discussion and to learn from it, then careful preparation is essential. It is likely that you will be given one or more specific topics that will form the starting point for the tutorial discussion. You will, therefore, need to read up on the topic and make succinct notes that you can use as prompts in the discussion.

If the tutorial topic is based on material that you have been covering in your lectures, then the obvious place for starting your reading will be your lecture notes, before moving on to read any specific references you may have been given. As a guide, the most important things to ensure are that:

- you are keeping focused on the question asked;
- you are making brief notes that summarise the key points and which you will be able to use as prompts in the discussion;
- you are keeping a record of the sources of the information (see Chapter 11, section 11.2.3 *Academic expectations*) so that you can cite the paper or text if you are challenged regarding the reliability of the information you are presenting, or if you want to find the information again to get more detail.

3.2.2 Contributing to academic tutorials

You have come to the tutorial having done your preparation. The next important element is to make sure that you participate. Participation, of course, means making active contributions to the discussions, but it also means listening carefully, thinking about what is being said, and relating that to what you have already learnt.

The first and most important thing to remember about tutorials is that they are not mini-lectures: your tutor will expect to have a role in facilitating and guiding the discussion but will not expect to be talking for much of the time. Indeed, some tutors are experts at sitting in silence waiting for the students in the group to start talking! There will, therefore, be an expectation that you will be an active participant—that you will attend the tutorial having prepared beforehand and be prepared to discuss your ideas, as well as to listen to the points made by your peers and the tutor.

You will probably find that this is all quite a daunting prospect at first and there is a strong temptation to sit looking at the floor, hoping not to be noticed. While this is an understandable response, it is not a very productive approach. Remember that your tutors are there to help you understand the subject and will want to help you develop your skills in communicating ideas; however, they can only do that if you participate. Don't be afraid of expressing ignorance if there is something that you don't understand—the only thing that is likely to be irritating to the staff is if it is apparent that you have not bothered to do any reading or preparation beforehand.

Depending on the structure of the tutorial, there may be some short presentations about topics that you and your peers have been asked to prepare in advance, followed by discussions. Alternatively, there may be specific questions that you have been asked to discuss, so the whole tutorial takes the form of a facilitated discussion. Irrespective of the format, there are key ways of engaging with the process.

- Listen carefully to any ideas being presented by the tutor or other students and note down the key points of the arguments. At the end of the statement, say whether you agree or disagree with the ideas, presenting evidence to support your views. It is no good just saying you agree or disagree with an idea—you also need to be able to justify that view.

- If there are things you don't understand, then ask for further details or explanation. Never leave the tutorial not understanding something because you were afraid to ask questions.

- At the end of the discussion, make sure you can summarise the key points for yourself.

- You do need to be sensitive, though, if you are a confident speaker, don't try to dominate the discussion—let other people express their views as well, since that is an important way of learning.

- Also, though, if someone else is dominating the discussion and you don't feel you have the chance to speak, then make sure you attract the tutor's attention, for example, by raising your hand.

3.3 **Practical classes**

The knowledge base in all areas of the Biosciences is underpinned by careful experimentation and observation. The ability to design experiments, to undertake them methodically, to observe and record the outcomes accurately, and to analyse and interpret the findings is vital to the development of a scientist. These skills, however, are not sufficient on their own: you also need to be able to communicate your findings to other people. If someone asks you for directions to find somewhere, knowing where to go is only part of the issue; you also need to be able to communicate that knowledge. In the scientific community, this is done in a number of ways, the most common being presentations at scientific meetings and published reports.

During your studies, you are likely to undertake a range of practical work, including practical laboratory work, fieldwork, and independent project work. The aims of these types of work are to develop your skills as a scientist and also to deepen your knowledge and understanding of the subject. The aims of this chapter are to help you get the most out of your practical work and, in Chapter 8, we will consider how to write up your practical reports.

Although the focus of this section is on laboratory-based practical classes and project work, much of the material is also directly relevant to fieldwork as well. To help create some context, the text uses as its examples an undergraduate investigation into the jumping ability of the locust. We will follow this theme through when considering the writing up.

3.3.1 **Before the practical class**

Your ability to produce a good practical report depends very much on whether you have undertaken the practical work effectively in the first place. If you turn up to the practical class not knowing what you are supposed to be doing and not having checked that you have brought everything you need with you, then you are certainly in danger of working less efficiently in the class and are also likely to perform less well. Good preparation beforehand is, therefore, a key stage to success in all types of practical work.

Much of this section will seem to be simple common sense, but if you want to get the most out of the practical work you are going to do, it is essential that you prepare properly. Not only will effective preparation help you learn more and get better marks, but it can often pay off in terms of saving you time and effort when you are actually doing the work. There are few things more frustrating than having spent a long time setting up an experiment only to discover that you have set it up incorrectly or not recorded all the data you need, so you have to start all over again.

Good preparation is particularly important if you are going to be working away from your normal site, for example, in the field, when you cannot easily go back to fetch something you have forgotten. More specifically, good preparation means knowing:

- where you have to be and when;
- what items you need to bring with you;
- what you are going to be doing and why;
- some background to the investigation;
- how you are expected to present your findings.

Read the schedule for the practical class: reading the schedule carefully will enable you to check the time and location of the class, and what items you need to take to the class. It is good practice to get into the habit of writing your experimental notes in your laptop or in a notebook. This will be particularly important for a research project, for which you will be expected to keep a lab record that you may need to submit along with your report. Most scientific employers will require you to keep detailed notes so that, if necessary, results can be verified later, so this is a skill worth developing now.

Read the details of the experiment: reading the details of the experiment will give you a good idea what you will be doing. It may well be that the techniques and equipment will be new to you, but familiarising yourself with the description of the procedures will help when it comes to doing the experiment. Highlight any key sections of the methods, e.g. if specific measurements need to be made at specific times. Figure 3.3 shows an example of how you

Figure 3.3 Using highlights to identify key parts of the experimental protocol

Jumping distance: experimental measurements
Measurements must be made in a warm environment (if possible 28–35°C).

You should try to obtain records from at least ten adult locusts. Each locust should be weighed and the length of the femur of the hindleg measured.

Take care not to mix up your locusts during the recordings: use ink-marks on the hindlegs for identification. The wings of the locusts should be fixed together with a small strip of adhesive tape to prevent them from flying.

might do this for an experiment. If you are going to be working in a group, getting together in advance to plan how you are going to organise yourselves and divide up the workload will help you to be more focused and save time.

Read up on the background: reading up on the background to the work will help you understand the biology involved and can put what you are doing into context. It will help to read through your notes from related lectures. If there is a reading list, try to read through at least some of the material beforehand and make notes of key points.

Check if there are instructions about presenting your results: checking how you should present your results is important because it may vary from class to class. Sometimes you will simply be required to fill in tables or draw graphs and answer some specific questions, but at other times, you will need to write a full report similar to a scientific paper (you will certainly be required to do this for a project report). Again, knowing in advance what is expected of you will help you to be more efficient in your work and save you time in the long run.

3.3.2 During the practical class

If you have done some preparation for the practical class, it is much more likely that the class itself will go well. There are two important things you need to do during the class to ensure you can write your report afterwards—make notes and record your results.

Make notes

Try to get into the habit of making notes during your practical classes. Many practical classes will begin with a briefing session addressing a variety of aspects of the work, such as safety issues and instructions, and demonstrations of the techniques to be used, as well as some background subject information. There will be other times during the practical class when it will be useful to make notes as well, for example, if you need to make changes to the procedures during the experiment or if you discuss some aspects of the experiment with your demonstrator or lecturer.

Before beginning the experiment, check through the sequence of tasks that you need to complete and, especially for timed measurements, make sure you have a clear list of what needs doing and when.

At the end of the class, you may well have another briefing session. Make the most of the information you are given because this could be very important for when you are writing up: make good notes—don't just rely on your memory, because you will forget things!

Record your results

Recording your results is clearly a critical part of undertaking any piece of practical work. You will probably record your experimental data directly into a spreadsheet or other program rather than on sheets of paper. However you are storing the information you still need to draw up tables for the results, making sure you identify what is being measured and what each column of the table represents.

During the experiment, record your results carefully, clearly identifying numerical values and the units in which they were measured. Also, make notes of any changes you made to

Figure 3.4 Two very different ways of recording data

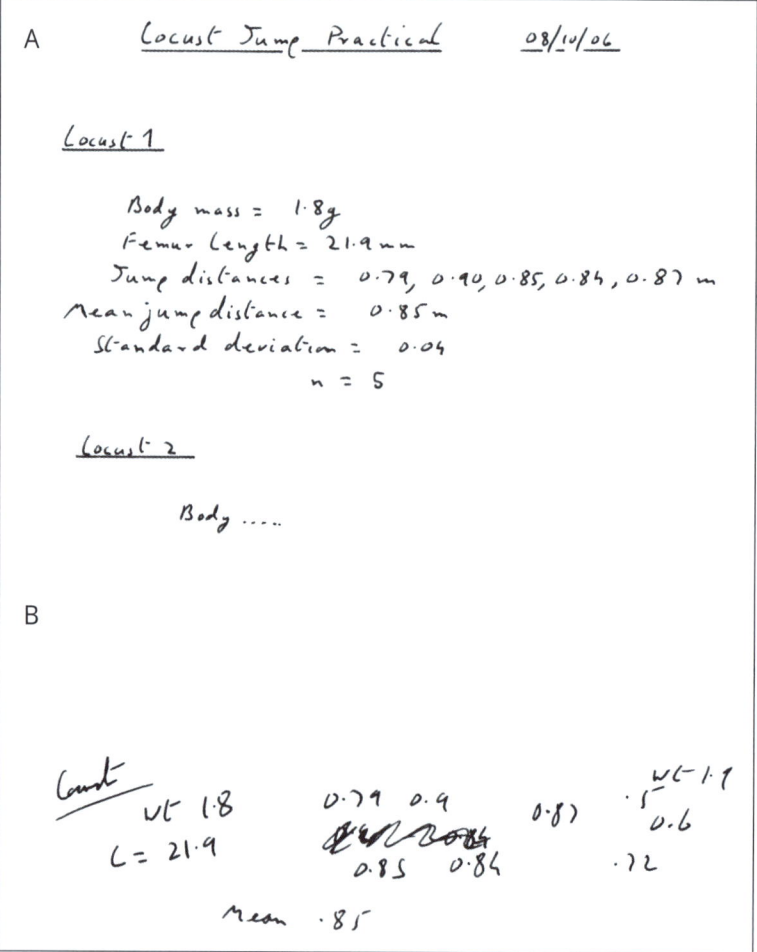

the protocol, or to the way in which the results were measured. Again, care taken at this stage will save you a lot of time and frustration later. You are also much less likely to make mistakes. Figure 3.4 illustrates two very different ways of recording the results from a simple practical class measuring the distance locusts can jump.

Compare the two sets of records in Figure 3.4 and list the aspects that you think are unclear in the data sets in sample B. If you were presented with these two data sets, would you be able to write about them in a meaningful way?

What if things go wrong?

By their very nature, experiments may go wrong, and you find you don't have any data or that they do not make sense. If this happens, speak to your tutor as soon as possible—it is likely that you will be able to obtain another set of results from your tutor or from your colleagues in the class that you can then use for your report.

At the end of the class

As soon as you have finished the experiment, there is often the temptation to pack off and leave as quickly as possible. Don't do this!

Make sure that you:

- take the opportunity to ask any questions of the demonstrators to clarify anything you don't understand;
- have a complete set of results;
- understand how to analyse and interpret your results;
- understand the principles of the experiment;
- know what is expected of you in terms of writing up the experiment and any deadlines for submission.

See Chapter 8, section 8.3 for guidance on writing up your reports.

 Chapter summary

In this chapter, we have considered some different forms of teaching sessions: lectures, tutorials, and practical classes. However, as we hope you have seen, there are some common themes to making the most of these sessions. Before the session, take some time to prepare by familiarising yourself with the objectives and the topic of the session. In particular, make sure that you read any materials that have been provided to you in advance by your lecturers. During the session, make sure that you keep engaged and take notes to remind you of the material. Finally, if there are any points that you are not sure of, or don't understand, do ask your lecturers.

4 What to do outside of taught sessions

→ Introduction

Even in subjects like the biosciences that have a strong emphasis on practical work and lectures, taught sessions are unlikely to take up more than half of an average week of study, so what do you do with the other half of your study time? In previous generations, students were said to go to university 'to read for a degree', the implication being that most of their working time as students would be spent reading up on the subject they were studying. A modern degree focuses on the concept of 'independent study', which might include tasks such as:

- investigation of the subject beyond the content covered in taught sessions (reading);
- complete prescribed tasks (assignments) to a set standard or brief;
- apply feedback to improve yourself and your work;
- reflect on your personal development and decide on future directions of growth;
- identify, troubleshoot, and resolve problems by finding and using appropriate sources of guidance and support (such as this book!).

A key thing separating university education from learning at school or college is that at university you spend most of your time doing 'independent' study (as opposed to guided study). This is because, a core part of what you are expected to develop and evidence by studying at university is the ability to effectively work independently. It is very important to note that 'independent' study does not mean you have to do everything alone. It is often helpful to have study groups to help you achieve your goals. There is also a raft of support available for students that need it (however the responsibility is often on you to access it).

4.1 Reading for a degree

Reading is going to comprise a very large amount of your independent study time. There are several reasons why you will need to read during your degree, and the approach that you take will depend on what your goal for the reading is. Some key reasons for reading include the following:

- reading specific texts set by the teaching team to prepare to engage with the degree in general or a specific taught session (such as reading a scientific paper to discuss during a tutorial);
- following up on taught sessions to reinforce and revise the subject matter you have been taught or to help you grasp concepts that you have found difficult to understand

(especially in instances where you expect to be examined on that knowledge—see Chapter 6 *Exams*);

- independently researching topics to develop a broader and/or deeper understanding of a topic required for an assignment (such as writing an essay or delivering a presentation—see Chapter 9 *Researching your coursework topic*);

- following up on topics that interest you, because through the library you have access to basically all of humanity's scientific knowledge.

This section provides a short overview of the library as a resource and a guide to the different types of text you will encounter during your studies. The aim of this section is to help you make the most of reading that has been prescribed, or set, by your lecturers (most likely in preparation for a taught session of some description). We will deal with reading focused on developing a breadth and depth of knowledge for your assessment in Stage 2 of this book. Specifically, there is more advice on reading for exams in Chapter 6, and detailed guidance on how to find, read, and incorporate literature into your coursework in Chapter 9. Help with critical reading and writing can be found in Chapter 12.

4.2 Making the most of the library

For many students, the library is the place where they spend the most time during their degree. It is often a quiet place, open long hours, well-heated and lit, and with access to all the essential digital tools required to effectively engage with the degree. It is free for students to access and often provides important community spaces. Many universities use their libraries as a central place for providing support with study skills. In short, for many students, the library is the best place to get work done.

4.2.1 What is a university library?

University libraries are not just a physical place but also an essential digital resource. Most of the textbooks and scientific papers that you need for your independent study will be available online. You should be able to use the library archives, and through them essential scientific research repositories, to research any subject you can think of. As outlined previously, throughout your studies, you will be expected to expand on what is taught on your degree through independent research. Only engaging with content covered during taught sessions is unlikely to result in achieving a good degree. So, the library (and its digital repositories) should be high on your list of places to be when not in a taught session, whether you are using a textbook to review the stages of a metabolic cycle in preparation for an exam, or investigating whether the reintroduction of wolves to the UK is feasible (or even ecologically desirable) to inform a piece of coursework.

Your university library will take up a considerable proportion of the tuition fees that you have paid; not least because access to scientific literature is expensive. Therefore, making the most of it is not just essential for doing well on the degree but a way of ensuring that you are getting value for money. We would recommend viewing your library as a similarly valuable resource to your lecturers, laboratories, and fieldtrips.

4.2.2 Navigating the contents of the library

The main reason that most people interact with their university library is to access the publications found within (or online through the university library portal). There is a wide range of types of publication available for use at different levels of study. Each of these has different characteristics and will probably be used in different ways. In this section, we will consider the main forms of publication, some of their key characteristics as study aids, and suggestions as to their use. This section is focused on explaining the different types of literary resource and how to read them, with the assumption that they have been set by your lecturers. Further consideration of how to effectively find literature and incorporate it into coursework is covered in Chapters 9 and 12.

A fairly simple way of categorising the types of publication you might want to use for your studies is as follows:

- textbook;
- specialist topic book/monograph;
- general science journal;
- specialist review article;
- research paper.

Let's consider each of these in turn.

Textbooks

The textbook is probably the form of publication with which you are most familiar from your studies at school or college. Textbooks still form a very important resource for students at university, predominantly in the first and second years of study. Whereas many of your textbooks will have been broad-based (for example, a textbook covering all of biology), the books you will be using, even in your first year of study, will be more specialised, focusing on a specific sub-topic of biology such as biochemistry, genetics, or physiology. The principle is the same, however: the textbooks offer a broad-based foundation to the study of a subject.

Buying textbooks

Even within the individual topics such as biochemistry or physiology, if you go to a bookshop or search online you will find a bewildering array of books available to buy. It is important to be cautious when purchasing books, otherwise you can find that you have spent a lot of money very quickly, sometimes on books that will not be of great use to you.

- Most universities publish lists of recommended books for specific modules or degrees, so start off by consulting the lists for the modules you are taking. Most universities purchase multiple copies of any essential or recommended books and have them available in the university library (in physical and digital formats). So, do not buy anything without checking the library first, you may well find that you have sufficient access without needing to purchase anything.
- Having decided you do want to buy a book, you have to choose which one. It may be that your reading list only suggests a single core text, which makes life easy. But there may be

a choice of recommended texts, perhaps covering the material to different levels and at different prices. Again, the library is a useful starting point. Have a look at the relevant texts available in the library catalogue: you will probably find that they are different in terms of their layout, the way diagrams are used, the level of detail, etc., so you may find it helpful to test-read parts of the book to find which style suits you best. In the library, you will find that some of these books will be available as hard copy, some as eBooks, some as both. You may also find that they differ in terms of the additional features they offer: for example, is there an associated website with more information or access to self-evaluation tests?

- Be selective: for example, is it worth buying a book for £40 or £50 for a subject area that you have to take as a single module in your first year, but which you do not intend to continue with later on in your degree?

- You may consider buying a second-hand copy of the book. Major online sellers are likely to have stocks of second-hand books for sale, especially if they are standard textbooks. Your university or Students' Union may also have a second-hand book shop or online alternative. The books should be cheaper than buying brand new so you could save money, but you need to be cautious. For example, make sure that it is the current edition of the book that is on sale. If the book has an associated website, make sure that you can still access it: many sites have access controls which make them unique to a single user and cannot be transferred if the book is sold on. Also, do check the shipping costs!

Using textbooks

Textbooks commonly provide a broad overview of a topic and can have several uses as study tools. Think about the reading goals that you have and that will guide you in the way to use your textbook. Some possible examples could be:
    As with any source of information, textbooks do have limitations, as follows:

- Because most textbooks offer a broad coverage of the subject (e.g. physiology, or ecology, etc.) they may well only provide a fairly superficial overview of a topic. This is important because it may be just what you need for giving you the background to the topic, or for helping you understand lecture materials in your first and second years. They almost certainly will not, however, provide enough detail for the later parts of your degree, or for researching material for coursework.

- Textbooks are often written by a small team of authors, who will not be specialists in all the topics covered in the textbook. As a result, the quality of the coverage of the different topics may vary. For some texts, the book is prepared by editors, with subject specialists writing the individual sections. These are more likely to have a more even quality of coverage of the topics.

- Textbooks do go out of date, and this is likely to happen faster than for your school or college textbooks. It can take several years for a textbook to be written and published, and then there may be a few years more before the book is revised and a further edition published. Although this may not cause problems when used for basic reading (since the key principles of a topic are unlikely to change significantly), if you require more specialist information, then it is more important that the book you are using is up to date. So, when you are scanning your library for textbooks, make sure that the books you select are not more than two or three years old.

Table 4.1 How to align reading textbooks to your study goals

	Example Goal	Suggested Action
1	I need to broaden my knowledge and understanding of a specific topic, for example the function of the kidney.	In a general physiology textbook, read through the chapter(s) on the kidney, checking on any learning outcomes identified in the text. Make abbreviated notes to aid your learning (see Chapter 3, section 3.1.5 on note-making strategies).
2	I have just had a lecture on the physiology of the kidney which focused on the loop of Henlé and I don't understand it!	Read through your lecture notes and any accompanying handouts. Look up the loop of Henlé in the index of your physiology textbook and read through the relevant section slowly. Then read it again, making notes about the organisation of the loop and the way it functions. It might help to go to the library and read about it in a couple of different textbooks. You should also ask the lecturer for help, but it is a good idea to try to think it through for yourself first.
3	I have an exam that will include kidney function, how do I use textbooks to help with my revision.	See Chapter 6, section 6.1.2 for guidance on active revision techniques. You can create notes from your lecture materials and then cross-check them with relevant content in the textbook. Modern textbooks also commonly have paper or digital revision aids, such as quiz questions at the end of each chapter, so make sure to make use of those.
4	I have just been given an essay to write on the regulation of ion balance by the kidney. Where do I begin?	Read through any relevant lecture notes and handouts first. Then read the sections in your physiology textbook that deal with ion balance, making detailed notes about each of the key ions in turn (e.g. Na+, K+, Cl−, etc.). Make sure you put the notes in your own words and include the source reference (see Chapter 9, section 9.4). Having got the general picture, you will probably then need to refer to more detailed sources such as a specialist topic book, review articles, or research papers (see Chapter 9 for guidance on this process).

Specialist topic books/monographs

As their name suggests, these are books that focus on a specific topic within the subject. These usually take three forms:

- An advanced textbook dealing specifically with a fairly narrow topic, such as kidney function. This will clearly be more detailed in approach than the broader textbook on physiology but will be written in a similar style. It will almost certainly have been written by a specialist, or group of specialists, within the field.

- A monograph usually takes the form of a long article or a short book on a specialist topic. This will typically be narrower in focus and at a more specialist level than the advanced textbook.

- At the research level, books are quite often based on the papers presented at a specifically themed conference. As with journal research papers, the individual papers may be specialised, being written for a readership of research scientists within the field. However, the grouping together of a whole series of papers on a specific theme may well provide you with a valuable starting point for your research for an essay topic. These books also frequently contain useful overviews of groups of papers in review form.

As with using textbooks, you need to make sure that the information you are looking at is up to date; indeed, the more specialised the book or research paper, the faster it is likely to go out of date. All of these, however, make valuable resources for advanced reading around a topic to increase your depth of knowledge, and they can also make very useful starting points in research for essays or similar assignments.

General science journals

There are several general science journals, such as New Scientist and Scientific American, that publish review-type articles aimed at the general interest reader with a scientific back-ground. These types of journal can be very useful in providing overviews of a specific topic and also for keeping you up to date with wider developments in science as a whole.

Specialist review articles

Many research journals include review articles as well as the papers describing the results of novel research. There are also numerous journals dedicated to reviews in the biosciences, such as the 'Annual Reviews in . . .' series or the 'Trends in . . .' series.

Recent review articles are a gold mine for students seeking information about a specific topic. These articles are typically written by a well-known specialist in the field who brings together, in a single article, a synthesis of a large number of individual research papers. The approach of these articles is usually designed to enable someone with some knowledge of the field, but not necessarily at a very detailed level, to derive a picture of the current ideas and developments within that field without having to read all the individual research papers.

Review articles are very useful for providing more detailed information about a topic, for example for supplementing lecture material, particularly at second- and third-year level. They also often make very useful starting points in research for coursework assignments, both because of the overview they provide and because of the references made to recent research work in the field. Your lecturers will almost certainly include some review articles in the reading lists for your modules, particularly in the later years of your degree programme.

Research papers

Research papers published in academic journals are the dynamic bedrock on which current knowledge is based. These papers are written by the research scientists describing their latest discoveries or developments within their field of research. In any honours degree programme, there will be an expectation that you will:

- read recent research papers in order to supplement the lecture material;
- use them in order to prepare your coursework assignments;
- use the information you have acquired from reading research papers to support the arguments you present in examination essays (this will be particularly true for finals examinations);
- critically evaluate research articles as part of your coursework (see Chapter 12).

As with the review articles, your lecturers will certainly include recommended research papers (some of which they have probably written themselves!) in the reading lists they give you to supplement the material covered in the lectures. The fact that someone who is an expert in the field has recommended a given paper suggests that the paper has merit as a piece of scientific writing and that the findings are likely to be reliable. In the same context, papers for publication in scientific journals are subjected to a peer review process: in this process, reviewers (usually two), who are also acknowledged experts in their field, are asked to read the paper and to comment on its scientific value, to rate its importance in the field, and suggest any amendments they think should be made. Although this is clearly not a foolproof process, and is certainly not an excuse for you to stop thinking critically when reading, it does give these papers a stamp of authority and reliability.

Research papers are the most specialised of all the standard sources of information and can therefore be quite intimidating and hard to tackle: for example, they will probably use terminologies and be based on ideas which you do not understand very well. Research papers are usually in a highly stylised format, presenting information in a way to which you are probably not accustomed. Finally, it must also be admitted that they don't always make the most interesting bedtime reading! So, if you are going to invest time and effort in reading research papers, you need to be very clear about your reading goals so that you get the most benefit from your input.

The structure of research papers

Most research papers are written in a standard format, comprising seven main elements.

- Title. The title is a short descriptor of the paper; as such it will give you a good idea of whether the subject matter is relevant to your needs.
- Abstract. The abstract is usually about 200 words in length and provides a short summary of the paper. In particular this will include the aims of the research, the key results, and a brief commentary about the conclusions. It may be that this is all you need to read in order to glean the key points, for example to support a specific point you want to make in an essay.
- Introduction. The introduction is the section of the paper where the authors set out the background to the study. This is often very useful both for helping you to understand the basis of the topic and the reason why the researchers undertook the investigation

they are describing. As such, it will normally include a useful summary of the preceding research and an analysis of the questions still to be answered. It usually culminates in setting out the aims of the study.

- Methods. The methods section describes the way in which the research was undertaken, including details of the experimental protocols and the procedures for analysing the data. Unlike the type of report that you might write, the protocols are often not described in full but rather refer to previous papers in which they have been described, for example:

The muscle preparation was prepared for electron microscopy according to the method of Saito and Jones (2000) . . .

This form of summary information can be fine for people working in the field who already have a good knowledge of the literature. For the student trying to understand the paper, however, this can be a nuisance because it means further searching out of papers in order to find the actual description of the method, particularly if, as in this case, the method is an old but standard technique.

- Results. In the results section, the authors will describe the results that they have obtained from their research and will also display them in different formats, for example as graphs, tables, or other illustrations. Where quantitative data are being presented, this will also be analysed using statistical techniques. Note that this section is descriptive and does not normally include any discussion of the results or reference to the findings of other researchers.

- Discussion. This section is where the authors interpret the findings of their research both in terms of the specific experiments they have undertaken and also in the context of the current hypotheses in the field, as developed by other researchers. This section should link back to the Introduction where the aims of the project were set out. Again, this section makes very useful reading as it links what has been reported in the paper to the existing research literature.

- References. This comprises a list of all the research papers, etc., that were referred to in the text of the paper. The reference list is often very useful in providing a source of further reading to help you fill in the background of your topic.

Reading a research paper

This is often a major challenge for many students. Research papers are aimed at readers who are researchers of the subject themselves. Research papers assume that the reader has a detailed understanding of the topic in question, the kinds of research activities and data common to that field, and experience navigating the format outlined above. They are certainly not written with students as the intended audience in the same way a textbook is. It sometimes seems that research papers are as impenetrable as possible on purpose, so that only people who are part of the 'in crowd' can extract the relevant information. While there has been a trend towards producing science communication that is aimed at a more general audience, reading papers is a skill that requires patience, practice, and some guidance.

So, what should you do now? We will revisit the subject of reading papers multiple times throughout this book. However, for now we will encourage you to remember the philosophy that underpins Chapter 2. Be strategic! Before you start reading the whole reading list for every module and every scientific paper you can find on every subject covered during the lectures, we recommend that you first understand what you really want to get from your reading. We provide some example goals and activities below:

Table 4.2 How to align reading scientific papers to your study goals

	Goal	Action
1	I have been given a literature review that provides background reading for the module/lecture, but I will not need to discuss it during the taught sessions.	If the recommendation is a literature review or monograph, it is likely there to help you understand what the main topic's perspectives in the subject are. Read through as much as you can and make notes of the names of any key theories or ideas that are discussed, as these will probably come up in the lectures. Keep your notes handy during the lecture so that you can understand the relationship between what you have read independently and what you are being taught. You might be expected to refer to some of this reading in your assessment, so make sure you understand how the contents feed into any exams or coursework you have on the module.
2	I have been given a scientific paper that reports on the results of a specific study as preparatory reading but I will not need to discuss it during the taught sessions.	Scientific papers are generally about a single experiment or related series of experiments testing a small number of very specific research questions. The most important points in most papers are the 'Key Finding' and the 'Research Questions', both of which should be in the paper abstract. Further guidance on how to quickly read papers and extract the most important points is covered in Chapter 12, section 12.2.3, so go there to learn how to do this. Keep your notes handy during the lecture so that you can understand the relationship between what you have read independently and what you are being taught. You might be expected to refer to some of this paper in your assessment, so make sure you understand how the contents feed into any exams or coursework you have on the module. For example, you might be expected to explain what one of the figures means in an MCQ at the end of the module.

	Goal	Action
3	I have been given a scientific paper that we will be discussing during an interactive taught session (such as a seminar or tutorial, see Chapter 3, section 3.2)	In this case, you are probably going to need to discuss the paper in some detail and maybe even critique it. Go to Chapter 12, sections 12.2.2 and 12.2.3 and follow the guidance on reading and critical evaluation of scientific papers. Bring your notes with you to the tutorial. Remember, this is a really difficult task, so just try your best and make good notes of your thoughts. If there are any steps that you cannot write the answer to, make note of them and contact the person leading the taught session (or ask about them during the taught session itself). The chances are there are lots of people who have struggled with this who are waiting for someone else to ask. Make the most of the opportunity to talk to someone who will absolutely know the answer to the question, because after that taught session is finished you may never get that opportunity again.
4	I am reading papers to independently research a topic in preparation for an assignment.	We will deal with this scenario in detail in the next part of the book (Stage 2 *Navigating assessment at university*), particularly in Chapter 9 *Researching your coursework topic*.

4.3 Independent study time and assessment

This chapter started with a question 'What do you do when you are not in taught sessions?', the simple answer is 'work on the actual things that will define your grades and the degree you graduate with'—the assessment. The taught sessions and supplementary training resources are where you learn the knowledge and skills required to do well on your degree. Completing the recommended reading activities is important to help develop breadth and depth of understanding. However, to actually pass the degree you must evidence that you have in fact met the learning outcomes: assessment is your opportunity to do this. You can attend all the taught sessions available but if you do not pass the assessment within the agreed timeframes then you cannot graduate with a degree.

It makes sense then that assessment should form the focus of your independent study activities, including your reading. Examples of how you should align your independent study activities to assessment are:

- The type and topic of your university assessment should inform what you read and what the purpose of the reading is (see Tables 4.1 and 4.2 earlier in the chapter).
- Whether you spend time revising taught content for an exam or researching a topic to expand on what you have been taught for your coursework.

- Whether you need to collaborate in structured groups on a group assignment (see Chapter 12, section 12.1), in informal study groups to support your navigation of the degree (for example, Chapter 6, section 6.1.4), or on your own to avoid potential collusion issues (see Chapter 9, section 9.5.2).

- When your deadlines are, and therefore when you need to be working and when you can allow an increased focus on aspects of your life external to the degree (Chapter 5).

- Guiding your personal reflective activities by 'making the most of your feedback' (Chapter 13).

Every university degree has a different assessment profile. If you worked through Chapter 2 *Know your course*, you should have a list of assessments to complete and the deadlines for getting them done during this semester (if you do not have a list, we recommend you go back and complete this activity). The majority of time you dedicate to completing your degree, that is not attending or preparing for taught sessions, should be spent planning, preparing, researching, completing, and reflecting on the assessments.

➕ Chapter summary

For many students, the most difficult aspect of university is not the complexity of the content being taught but the level of responsibility students must take for their time and what they do with it—also known as 'independent study'. This expectation is often not made very explicit to students, and it can take a while to get used to. This chapter has attempted to provide some guidance on how to manage the level of independence and autonomy that comes with studying for a degree at university. We have covered some ideas of what you might do with your time other than attend as many taught sessions as you can. We have included some guidance on reading, what to read, where to find it, and what to look for when reading, which is likely to take up a very significant amount of time on most Bioscience degrees. Finally, we have attempted to impress upon you how assessment should help you focus and direct your independent study time, so that it is strategic and purposeful. The next part of the book is 'Stage 2 *Navigating assessment at university*', which provides further guidance on that aspect of your degree.

Stage 2

····································

Navigating assessment at university

····································

Assessment is how you evidence that you have met the learning outcomes of the degree. You can attend all the taught sessions available but if you do not complete and pass the course assessment (within the agreed timeframes) then you cannot graduate. Stage 1 focused on how to maximise learning, but Stage 2 is about ensuring that you can effectively evidence that learning to your lecturers.

Assessment varies across degree subjects and universities; your degree may be heavily exam-based or use a variety of coursework formats such as essays, posters, and presentations. Different assessments necessitate different learning strategies, so you need to adapt your engagement with the taught aspects of the course depending on the assessment type. It is important that you know which assessments you need to complete and when (we recommend having a deadline calendar as outlined in Chapter 2). This variety means that we have to cover a wide range of assessment types within the book, not all of which will apply to you. However, we have attempted to provide some unified approaches to assessment that you can apply beyond the exact examples included herein.

Remember, it is up to you to work out what assessments you must complete and the best way to manage your time to meet multiple complex deadlines (just as it would be in any other professional environment). So, Chapter 5 starts the stage by giving you an overview of broad strategies you can use to organise yourself, which can apply to all the assignments you are set as well as your personal and professional lives. Chapter 6 covers exams in their entirety from revision strategies, different exam types, how to answer questions, and how to deal with exam anxiety. The rest of Stage 2 is dedicated to the coursework creation process, which is broken down into:

- Chapter 7 *Preparing for coursework assignments* focuses on how to work out what you are being asked to do by your lecturers and overcoming common barriers to the creative process.

- Chapter 8 *Creating written coursework* provides guidance on how to plan and organise content for the two most common aims of Bioscience assignments: (1) independent discussion of a topic (e.g. essays) and (2) reports on data (e.g. scientific reports).

- Chapter 9 *Researching your coursework topic* outlines strategies for effective independent investigation of the scientific literature, which is an essential aspect of doing well on every coursework assessment. The chapter also considers the importance of building academic integrity and avoiding academic misconduct.

- Chapter 10 *Developing visual and oral presentation skills* covers techniques on visual design and public speaking for scientists, helping you to communicate in a wide variety of formats during and after the degree.

- Chapter 11 *Revising drafts and finishing touches* is all about turning coursework drafts into polished final submissions, including some guidance on common professional expectations such as grammar and reference lists.

- Chapter 12 *Key competencies: collaboration and critical thinking* covers group work and critical thinking. These skills are rarely taught directly in the degree but are of great value to both your studies and your future employment. This chapter will help to ensure that you can work professionally on group assignments and evidence independent critical thought in your coursework.

5 Getting yourself organised

→ Introduction

In Chapter 1, we introduced the idea of managing your time well as a foundational skill. We also said that it's one of those things that is much easier to say than to do; in theory it's straightforward, in practice it's difficult. But this doesn't just apply to managing your time, it applies to organising yourself and your life more generally. In fact, a significant proportion of functioning well as an adult is about being able to do things by a certain time (essays or tax returns) or finding things when you need them (that useful reference or where you put your passport).

As far as your academic work is concerned, the chances are, you won't get chased when a deadline is looming, you may not even be told that you've missed it, you'll just find out that you've failed that piece of work.

However, someone telling you how to organise yourself or your time can be a bit annoying. It's a bit like someone telling you that you need to do more exercise, even if they're right, being told doesn't usually help. But the solution is similar too, getting yourself organised, as with exercise, is quite a personal thing. There are some principles, of course, but those principles won't be applied unless you find something that you can build into your day and that works for you (as with exercise).

A good test of any system is whether it works when things get difficult. It's easy to be smug about your natural ability or the habits you've developed, but do they work when the deadlines pile up or unforeseen circumstances occur?

We'll use this chapter to identify some useful principles and tools and think about how to apply these to your studies.

5.1 Why you need a system

You might not think you need a system to organise your time or yourself, but you do. Studying at university is a lot more autonomous than studying at school or college, and there are far more distractions, either because there's lots going on socially, or because you have more responsibilities (part-time work, looking after yourself, or even caring for others). With greater autonomy needs to come greater responsibility, and one way to take responsibility is to develop methods or habits (what we're collectively referring to as a 'system') that helps make that responsibility easier.

What do we mean by 'system'? The word 'system' might make you think of something that's sophisticated or complicated, but it needn't be. In fact, the best systems are simple. Remember, in the context of organising yourself, what you need from your system is something that provides the structure you need to function well as the pressure increases (multiple deadlines, unforeseen circumstances). What you're trying to do is reduce your cognitive load and associated stress levels by creating a system (or systems) that takes the strain, rather than you.

We're going to think of systems in two parts; what the system is for (the principles) and how you can implement it (the tools).

5.2 Principles to help you get organised

When it comes to getting yourself organised, it can be helpful to think in terms of projects. Whether it's big or small, whether it's writing an essay or going on holiday, most things can be thought of as a project. A project can be defined as:

- a set of tasks;
- that need to be completed within a defined timeline;
- to achieve a specific set of goals.

Let's use these points to identify some principles.

5.2.1 Identify what you're trying to achieve

Before you dive into the doing, at the start any project you need to identify what it is you're trying to achieve. This is aways the first step; without a clear aim, you can't break the project into smaller tasks and you're likely to do things either in the wrong order or waste your time on doing things that don't help you achieve your goals. The bigger a project is, the more important it is that you develop a clear and simple statement of what you're trying to achieve, for example:

- write a lab report;
- get a part-time job;
- complete a parkrun.

Once you have a clear statement of what you're trying to achieve you can create a specific set of goals, for example:

Write a lab report:

- complete on time;
- fit it around other commitments;
- improve on my last mark.

Get a part-time job:

- find something that fits around my studies;
- draft a CV and covering letter;
- ask someone to check it before I submit.

Complete a parkrun:

- in two months' time;
- without getting injured;
- and have fun!

Once you've identified the specific goals you want to achieve you can break it down into tasks.

5.2.2 Break it down into tasks

Let's take the parkrun as an example. A parkrun is a free, community event where you can walk, jog, or run a 5-kilometre course in your local area. They take place every Saturday morning and are designed to be positive, welcoming, and inclusive. We'll assume that you've never done a 5k before and you need to do a bit of training. So, how do we break this down into tasks? The tasks will relate to the goals, so use these as your starting point, in this case: how do you complete a parkrun in two months' time, without getting injured, and whilst having fun? It's helpful to draw up a list of tasks; don't worry about the order or the relative size of each task, just get them down.

- Find some appropriate shoes;
- Figure out how far can I run already;
- Decide if I want to do it on my own or with someone else;
- Find a training partner;
- Find a local parkrun;
- Register for parkrun;
- Decide if I want to run it or walk it;
- Find suitable clothes;
- Find a training plan;
- Schedule training sessions.

Once the tasks have been identified they can be grouped and sequenced, for example:
Training:

- Find a training partner;
- Find a training plan;
- Schedule training sessions.

Admin:

- Find a local parkrun;
- Register for parkrun.

Practicalities:

- Find some appropriate shoes;
- Find suitable clothes.

Questions:

- Do I want to do it on my own or with someone else?
- Do I want to run it or walk it?
- How far can I run already?

Once you've broken down your project into tasks, you're ready for the next stage; deciding what needs to be done by when.

5.2.3 Decide what needs doing by when

A list of tasks without a clear sense of what needs doing when can be overwhelming. Without putting tasks into a timeframe, they're unlikely to get completed on time and you're unlikely to achieve your goals. You therefore need to decide what needs doing by when. Later in the chapter, we'll look at the specific examples of exams, coursework, and research, but in relation to our parkrun example it will include:

- identifying your fixed points over the next two months (coursework deadlines, social commitments, etc.);
- if having identified your fixed points you think you can't fit in the tasks to achieve your specific goals, decide what's flexible and what isn't;
- organise your tasks appropriately around your fixed points, probably by using a plan or calendar;
- review your progress on a regular basis to keep yourself on track.

5.2.4 Take the next step

Of course, there's much more that could be said about projects: dependencies, risks, resources, decision making, and so on, but as we've said, it's helpful to keep things simple. One more thing that is always worth thinking about, however, is 'what's the next step?'. Projects often grind to a halt due to inertia related to the next step, even when they've been carefully planned and sequenced. Once you have a plan in place, whilst it's good to understand the big picture and discern your progress against the whole, what's often needed is simply to take the next step.

5.3 Tools to help you get organised

For some of you, this section is going to seem obvious; this could be because you're naturally quite an organised person, or because you've already got tried and tested tools to keep yourself organised. But remember, the more deadlines mount up and unforeseen circumstances occur (which will happen), the more you're going to need a system, so scan through what follows to check that your confidence is well founded. If what follows doesn't seem obvious, or perhaps is obvious but you know that you could improve in this area, you should read this section carefully.

5.3.1 Find places to put things

When you don't have many things, or the things that you have aren't critical, or you can rely on others to help you, having places to put things won't seem very important. However, at university you'll have more things (because there's more going on), the things are more critical (a module handbook or your tenancy agreement), and you'll be expected to manage these things yourself. The best way to tackle this is to have a system. It's helpful to consider both digital and physical systems.

Digital

Increasingly, things are only available digitally, in fact most administration has gone paper-less. In theory, this makes things more accessible and searchable, but as the volume of information you have increases, it gets more difficult to find what you need. Obviously, you can do a search of your emails or digital files, but that may not always work because it may yield too many results to be useful. You therefore need to think about imposing some order on your digital content. For files (documents, spreadsheets, etc.), this could be a series of folders and sub-folders (perhaps organised by module); for emails, you could consider labels or folders; for other digital information, you could think about bookmarking. There are many online tools that can help, including:

- Google Drive or Microsoft OneDrive for files;
- Google Keep, Microsoft OneNote, or Evernote for notes and images;
- Browser-based bookmarking or favouriting for other online information.

What's more, each of these tools can be used collaboratively, making it easier to work with others on shared goals.

> ### Try this—Share ideas about digital filing
>
> The suggestions above, whilst a good starting point, are quite limited. You need to find tools that work for you. Ask other students what tools they use to organise digital content and share your approaches too. See if you can help each other discover different methods, considering the pros and cons of each.

Physical

Fewer and fewer things are only available in hard copy only (passport, driving license), but that doesn't mean you don't need a system. If you have multiple hard copy items that you need to keep safe, then a filing box or folder is useful; if you only have a handful of items, then simply a safe place that you always use will be sufficient, for example, a bedside drawer. The trick is to get into the habit of always filing these important things in the correct place, and if you use them, put them back in the same place when you've finished. This is obvious, but it saves a considerable amount of stress when you can locate what you need quickly and easily.

5.3.2 Make to-do lists

There are many ways of managing tasks; it's about finding a system (using tools, methods, or habits) that works for you. A good task system (whether physical or digital) needs to be able to:

- Store—does it have sufficient capacity to contain all the tasks I want to keep in it?
- Categorise—does it allow me to categorise tasks that belong together?
- Retrieve—can I easily retrieve the information stored in it?
- Remind—can it remind me when (and perhaps where) I should complete the task?
- Connect—can I easily connect information (e.g. from emails to tasks)?

Digital

There are some excellent digital tools that can help you manage tasks (performing all the functions listed above). One of the advantages of a digital tool is that you can access your task lists from multiple places, wherever you're logged in (on any device, whether via an app or a browser). They include:

- Google Tasks;
- Microsoft To Do;
- Apple Reminders;
- Remember the Milk.

Physical

If you choose to use a physical system, for example a carefully curated notebook, you'll need to develop some effective habits alongside so it can do everything you need it to do. A physical system isn't going to be able to connect easily to information elsewhere, but it certainly has the capacity to store information; with careful use of colour or space, it can also categorise; labelling or indexing will help you retrieve information, and with regular reviewing it can even be used as something to remind you when tasks need to be done.

> ### Try this—Share ideas about how you manage tasks
>
> The suggestions above, whilst a good starting point, are quite limited. You need to find tools that work for you. Ask other students what tools they use to manage tasks and share your approaches too. See if you can help each other discover different methods, considering the pros and cons of each.

5.3.3 **Use a calendar**

Let's do a quick recap:

- if information needs to be stored so it can be found later, put it in your filing system (or other specific location) for reference;
- if the information isn't for reference and doesn't relate to you being in a certain place at a certain time, but is something you need to do, it's a task;
- if information relates to you needing to be in a certain place at a certain time, put it in your calendar.

The busier your life gets, the more important a calendar becomes. Whether a calendar is physical or digital is up to you, but if you choose physical, you'll be going against the grain and will need to work harder at keeping it up to date. You may also choose to use different calendars for different things (e.g. work and leisure), but if you do this, you'll need to remember to check how one impacts on the other (e.g. is it sensible to have a late night tonight if I've got a 9 am lecture tomorrow?)

One of the things a calendar enables you to achieve is not just to turn up at the right place at the right time, but also you have less to think about (your cognitive load) because you know that you can rely on your calendar (your 'system') to tell you where you need to be when.

In setting up any calendar system, here are a few things for you to consider:

- Have I got it when I need it?
- Is it up to date?
- Does it notify me?
- Can it help me find related notes?
- Can it tell me where I need to be?

Digital

If you use an online calendar linked to your phone (probably via an app), then the answer to all the above questions can be 'yes'. That doesn't mean it's a perfect system, but it can certainly do everything you need it to do.

Physical

If you use a paper-based calendar, then, as with a physical to-do list, you'll need to develop some effective habits alongside so it can do everything you need it to do. A slim-line diary would fit in your pocket (they're about phone-size) so you could have it when you need it. Keeping it up to date will involve manually making sure appointments are entered correctly and changed if required. It won't give you a push notification, but you can just check it each morning (this is what people used to do in the old days!) If you choose to use a carefully curated notebook as a physical to-do list, you could combine a to-do list and calendar in one physical system, making it more likely you can find related notes. And assuming you keep it up to date, it can tell you where you need to be.

BOX 5.1 A note about phones

Phones can be a really useful tool to help you get more organised, but they can be a huge distraction too. If you are going to use your phone to manage files, appointments, or tasks, you need to be aware of the downsides. Phones, whilst useful, are not only distracting, many of the apps and features are designed to be addictive. Using a laptop to help you get organised is far less distracting or addictive simply because you don't have it in your pocket the whole time, but for accessing information when you want it, or making yourself a reminder when a thought occurs to you, a phone is much more convenient. So, how can you make the most of your phone without it becoming a distraction (usually caused by social media) or addictive (a compulsion to check your phone frequently)? There is a lot of good advice available on 'digital minimalism' or 'how to break up with your phone', if either of those sound interesting or useful to you, look them up. However, here are a few things that will help. Doing all of them at once would probably be difficult and possibly overwhelming, so pick one or two that you'd like to try and see if it helps.

1. understand if there's a problem—look at your phone's settings (*Digital Wellbeing* on an Android or *Screen Time* on an iPhone) to see how much time a day you're spending on your phone;

2. delete the apps you don't need (try accessing social media through your browser rather than the relevant app, it will feel a bit clunky but it's less addictive);

3. minimise notifications—turn off all but the ones most necessary (especially on the lock screen);

4. get an alarm clock that isn't your phone;

5. give your phone a 'bedtime' (try turning it off an hour before you go to bed);

6. try not to turn on your phone as soon as you wake up (try turning it on after breakfast);

7. when you're waiting for something (waiting for a bus or queuing in a supermarket) try and do something other than be on your phone.

None of these actions on their own will transform your phone usage, but they will help you reflect on your relationship with your phone and help you consider whether you might want to change it.

5.4 Applying these principles and tools to your studies

Earlier in the chapter, in section 5.2 *Principles to help you get organised*, we said that most things can be thought of as projects and that a project can be thought of as a set of tasks, that need to be completed within a defined timeline, to achieve a specific set of goals. In section 5.3 *Tools to help you get organised*, we considered a range of tools that can help you find things, do things, and schedule things. Now let's think about how this applies to your studies.

Preparing for and sitting exams, creating coursework assignments, and researching a coursework topic can all be thought of as projects. So, for each of these areas, we're going to use the project structure (a set of tasks, that need to be completed within a defined timeline, to achieve a specific set of goals) to help you get more organised.

5.4.1 Exams

Being able to organise yourself is a significant factor in doing well in exams. In Chapter 6 *Exams*, we cover a wide range of material including active revision techniques, practising outputting the information you have learnt, and what to do in the exam room. Here, however, the focus is on getting yourself organised for revision.

Identify what you're trying to achieve

There are a range of specific goals you could set out to achieve for exams, depending on whether the exams are formative or summative, and whether they count towards your final mark. For instance, for formative tests at the end of semester one in your first year, your goals might simply be:

- familiarise myself with university-level tests;
- learn as much as I can from the process;
- achieve a pass mark that I'm happy enough with.

Whereas for summative exams in your final year your goals might be:

- finish my dissertation sufficiently early to have enough time to revise;
- create and follow a revision timetable that gets me ready;
- achieve a pass mark that I'm happy with.

The context has a bearing on the specific goals you set out to achieve and it's important at the beginning of any 'project' to take a step back and think about what it is you want to achieve, rather than just diving into the tasks.

Break it down into tasks

Let's use the example of the goals from the summative assessment in your final year. As per the advice in section 5.2.2 *Break it down into tasks*, this might include the following (note, these are all covered in detail in Chapter 6):

- Check I have all the necessary material;
- Decide how I'm going to allocate my time;
- Finish dissertation by the second week of the Easter vacation;
- Check that my revision environment is suitable;
- Find out what I'll be examined on;
- Note how much time I have before the exams.

Once you have a list of tasks, group and sequence them (this may also prompt other tasks that you realise need doing).

Dependency:

- Finish dissertation by the second week of the Spring vacation.

Content:

- Find out what I'll be examined on;
- Check I have all the necessary material.

Scheduling:

- Note how much time I have before the exams;
- Decide how I'm going to allocate my time.

Environment:

- Check that my revision environment is suitable.

Decide what needs doing by when

You're now ready to schedule the tasks. Remember the other factors we mentioned earlier:

- identify your fixed points;
- decide which of your fixed points are genuinely fixed and which could be flexible;
- organise your tasks appropriately around your fixed points (using a plan or calendar);
- review your progress on a regular basis to keep yourself on track.

5.4.2 Coursework

Performing well in coursework is at least partly dependent on being able to organise yourself. In Chapter 7 *Preparing for coursework assignments*, we cover a range of areas including analysing the brief, drafting the material, and reviewing and redrafting. Here, however, the focus is on getting yourself organised.

Again, we'll use the project approach to highlight the important considerations.

Identify what you're trying to achieve

If you leave your coursework to the last minute, your specific goal might simply become 'hand in coursework before the deadline to avoid getting penalised'. However, with a bit of organisation you can be much more ambitious, this might include:

- produce a piece of coursework that I'm happy with;
- improve on my mark from my last piece of coursework;
- hand it in on time without needing to do lots of work the night before.

You're now ready to break your goals into tasks.

Break it down into tasks

In Chapter 7 *Preparing for coursework assignments*, we use the following structure, which we'll use to shortcut the brainstorming process here.

- analyse the brief;
- research the topic;
- make a plan;
- write first draft;
- review and redraft;
- proofread.

These tasks are broadly applicable, whatever type of coursework you're dealing with, including essays, lab reports, or presentations.

Figure 5.1 Planning your time for an assignment over the coming month

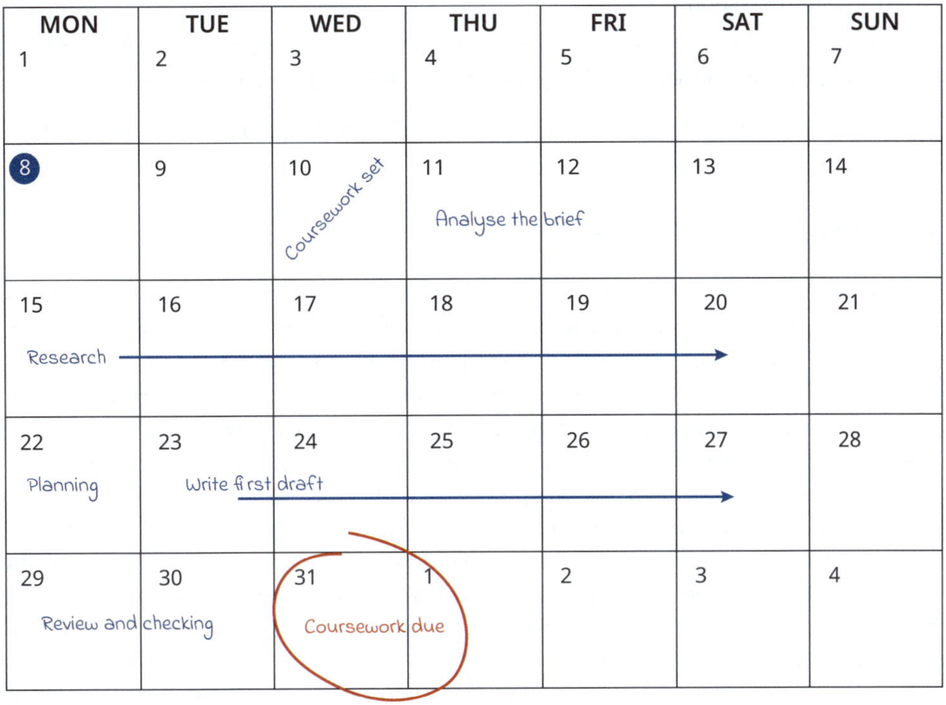

MON	TUE	WED	THU	FRI	SAT	SUN
1	2	3	4	5	6	7
8	9	10 *Coursework set*	11 *Analyse the brief*	12	13	14
15 *Research*	16	17	18	19	20	21
22 *Planning*	23 *Write first draft*	24	25	26	27	28
29 *Review and checking*	30	31 *Coursework due*	1	2	3	4

Decide what needs doing by when

Now the project is broken down into tasks, we can decide what needs doing when. Let's imagine the coursework is set three weeks before the completion deadline. Figure 5.1 shows a suggested schedule for completing the coursework on a schedule that not only allows you to not rush it the night before, but also produce a piece of coursework that you're happy with that improves on your last coursework mark.

5.4.3 Research

We're sure you've got the idea of the project management approach: identify what you're trying to achieve, break it down into tasks, then decide what needs doing by when (and don't forget to keep taking the next step). Rather than repeat those principles again for getting yourself organised in relation to research, we're going to focus on the tools.

In Chapter 9 *Researching your coursework topic*, we cover sources, note-taking, academic integrity, and plagiarism. Let's consider plagiarism briefly, because plagiarism is often caused simply by poor organisation, either of your notes, your time, or both. The principles to help you get organised (section 5.2) will certainly help you avoid plagiarism, but so will the tools (section 5.3); tools that help you to find things, do things, and schedule things (using filing systems, to-do lists, and calendars). So, whether you are using some of the tools referred to in section 5.3, or other tools you have discovered, they can help you:

- organise your notes (using physical or digital systems) so you quickly locate sources and their references when you need them;

- manage your tasks by using to-do lists (physical or digital) to help you keep on top of what needs to be done;
- plan your schedule by using a calendar (physical or digital) to help you keep track of what needs doing by when so that you don't rush things at the last minute.

The result of using these tools will be that it is much less likely that you will plagiarise.

5.5 Getting back on track

Finding a system that works for you is essential to both getting and keeping yourself organised, but even when you find something that works, sticking to it can sometimes be difficult. It could be that you begin to struggle to find things when you want them, or items just sit in your tasks list not done, or your calendar isn't working for you for some reason. If this happens, you need to stop and think, is it the system, is it you, or is it the combination of both?

Any system, whether physical or digital, and whether it's helping you with your filing, tasks, or scheduling, will need reviewing every now and again. What's working well? What isn't working well? Why are the bits that aren't working well not working? Spending a bit of time answering these questions (perhaps with the help of someone whose system is working well for them) will help you adapt your system to get it working for you again, or perhaps prompt you to try something new. Remember, you're trying to let the systems take more of the strain so that you don't have to.

Perhaps there's nothing wrong with how you've set up your systems, but you've just fallen into some bad habits (not filing things, not working through your to-do list, not reviewing your schedule). If this is the case, then you need to figure out how to get back into good habits. This isn't easy, especially if you're feeling overwhelmed, so take things one step at a time, think about the progress you made previously, and pick one thing that you can do to improve things and start with that. Reflect on how getting back on track with one small thing feels positive, then pick another. As we said, it's often not easy, but taking things one step at a time is usually the best way.

➕ Chapter summary

Some people are naturally good at organising themselves, others aren't. But even for those people who are naturally organised, when things get busy or unforeseen circumstances occur, we can start to unravel. Whether the principles and tools covered int his chapter were obvious to you or not, we hope that it's been useful. Remember, a good system is one that'll still work when things get difficult. However, even a good system can be overstretched. It's worth reminding ourselves what a good system helps achieve; a good system doesn't just help us get more and more done, it takes the strain so that we don't have to. A good system, therefore, is not about enabling us to be increasingly productive, progressively getting more and more done. That would lead to burn out, just at a later point than it would have happened without a system. Rather, a good system helps us not just to be productive but also helps us live more balanced lives, making time not just for getting things done but also for relaxation and enjoyment.

6 Exams

→ Introduction

Not many people like exams; this is partly because there's a lot riding on them, and partly because they can be difficult to prepare for and so end up being rather stressful. We can't do much about the former, but we can help with the latter.

Broadly speaking there are two types of assessment: formative and summative. Formative assessment helps guide further learning; summative assessment evaluates learning at the end of a period of study. Summative assessment can be either course-work (practical reports, presentations, essays, etc.) or, of course, tests or exams. How you perform in exams will have a large impact on your degree classification, because for some courses, exams account for a significant amount of your final marks.

Depending on when you are reading this book, exams might seem like a long way off, so why have we put a chapter on exams (that generally happen at the end of a course) near the beginning of the book? The reason is this (and this applies to all forms of summative assessment): knowing how you will be assessed at the end helps you know where and how you should focus your attention and effort now (see Chapter 2 *Know your course*).

We have divided this chapter into three sections:

- revising for exams;
- just before exams;
- sitting exams.

Section 6.1 *Revising for exams* covers getting yourself organised, using active revision techniques, seeing the big picture, and practising outputting the information you have learnt.

Section 6.2 *Just before exams* covers the wrong and right sort of last-minute work, and other last-minute preparations.

The final section 6.3 *Sitting exams* will help you to think through what you need to do when you're in the exam room; everything from getting settled to answering the different types of questions you might be faced with.

Remember: unlike most other topics covered earlier, you won't have this book to hand for guidance when you're actually experiencing the topic of the chapter—that is, when you are sitting in the exam room. So, it is worth spending time reading carefully to make sure you remember all the key points, so that you have them in mind and ready to put into action when exam time comes.

6.1 Revising for exams

One important way of alleviating some of the stress and increasing your chances of doing well is to get the revision part right. If you can go into an exam confident that you have prepared yourself as well as you could have done (or, at least, reasonably well), then a significant amount of the pressure can be eased, making it more likely that you can perform well. However, getting the revision part right takes a significant amount of organising; it is no good leaving it to the last minute. Revising for exams is a bit like training for a marathon: if you want to do well you can't cram all your training into the last few days—you've got to make a sensible training plan for the weeks and months beforehand and then do your best to stick to it.

Firstly, let's consider what revision actually is. In very general terms, revision is to do with going over a subject again, usually in preparation for some kind of exam. But exactly how you go over a subject again is critical to whether or not the revision is effective. You will know if your revision is effective because revision that is effective will make the material you are revising easier to:

- understand;
- remember;
- apply.

6.1.1 Get yourself organised!

We have already highlighted that getting revision right takes a significant amount of organising. However, getting yourself organised is easy to say but is more difficult to do, especially when you're beginning to feel the pressure of exams looming and you just want to get on with revising. It's important, however, to see the organising or planning aspects of revision as part of the process, rather than as something to get out of the way so you can get on with the 'real work'. Time spent organising revision is time well spent (as long as you don't use it as an excuse to procrastinate!)

There are several questions you need to ask yourself to get organised.

- What will I be examined on?
- Do I have all the necessary material?
- How much time do I have before the exams?
- How am I going to allocate my time?
- Is my study space appropriate?
- Do I know where I can go for support?

These questions will form the structure of this first section.

Find out what you will be examined on

Firstly, then, you need to find out as much as possible about what you will be examined on. This includes both the format and scope of the exam (or exams). In terms of format, this would include finding out answers to the following questions.

- How long will the exam last?
- How many questions will there be?
- Where will the exam be and how will I complete it (e.g. invigilated, in-person, written exam versus at-home online exam)?
- Will I have any choice about which questions I answer?
- What type of questions will I be asked (e.g. multiple choice, short answer, or essay)?
- What kinds of material will I be allowed to bring with me (e.g. nothing [closed exam], prepared essay plans [seen exam], or prepared notes and lecture materials [open exam])?
- Do I need to have done additional reading beyond the lecture materials?

Knowing the answers to these basic questions will help you to know what to expect. The better informed you are, the better you can prepare. It also means that you're not surprised when you open the exam paper and so will be more able to concentrate on answering the questions rather than spending time trying to understand issues of format.

In addition to the format of the exam you also need to consider its scope. What are you expected to know in terms of both breadth (the range of information) and depth (the amount of detail)? Imagine you were revising for an exam on comparative animal physiology: if you had revised enough in terms of breadth but had revised insufficient depth, your knowledge of comparative animal physiology would be too shallow. For example, you may have revised the comparative physiology of the respiratory, cardiovascular, and excretory systems, ion balance, and vision, but failed to learn them in sufficient detail. Alternatively, if you had revised enough in terms of depth but had revised insufficient breadth, your knowledge of comparative animal physiology would be too narrow. For example, you may have revised vision in intricate detail, but if you failed to learn about the other aspects as well (respiratory, cardiovascular, and excretory systems, and ion balance), then you would be very limited in the number of questions you could answer.

This may all seem fairly obvious, but finding out what you will be examined on is an important and often overlooked area of revision. There are a number of ways you can find out information about exam format and scope including:

- learning outcomes;
- module handbooks;
- past papers;
- course tutors.

Each of these will give you important insights into what you can expect.

Check you have all the necessary material

The second question to ask to get yourself organised is 'Do I have all the necessary material?' When you have found out about what you will be examined on (the breadth and depth) you then need to look at the material you have and decide whether you have enough (and whether you are expected to have anything specific prepared for the exam such as an essay plan for seen questions). It's important that you think about more than just your lecture notes at this point. Other important sources of information include:

- notes from tutorials;
- essays you have written for coursework;
- lab reports you have submitted;
- notes from presentations you have been required to give;
- notes from additional reading for all of the above;
- feedback from tutors on all of the above.

As you can see, the amount of information you will have accumulated over a term or semester will be considerable. At this stage, you are not supplementing your notes, simply auditing them to check where there are any gaps. We said earlier that module handbooks were an important source of information on exam format and scope, and as such, they help you to identify gaps in your notes and so help you determine whether you have enough material. Simply make a list of what information you have on a summary sheet and compare it to what you are expected to know, as described in the learning outcomes for the particular module you are revising for.

When you identify gaps in your material (we say 'when' rather than 'if' because it is unlikely that you will have everything), make a note of what the gaps are so you can come back to them later. It's also worth identifying whether the gap is a breadth issue (an insufficient range of information) or a depth issue (insufficient detail). The other factor to consider is whether certain gaps actually matter: if the format of the exam allows you some choice of which questions you answer (e.g. some essay-based exams) you may not need to revise everything. It is usually better to have revised most things in sufficient depth rather than everything at just a surface level. It may also be that there were some sections of the module or course that you just could not get to grips with and still don't understand. There are risks associated with this strategy of selective revision, so you need to be careful. For essay-based exams where you have a choice of questions, it's still very important that you revise more topics than you actually have to answer questions on, so that you do have a choice when you come to read the paper. A reasonable rule of thumb is always to learn at least two more topics than you need answer questions on, so if, for example, your typical essay paper requires you to answer three questions, then it is best to know five topics really well.

Note how much time you have before the exams

The third aspect of getting yourself organised is to be clear about how much time you have to revise. Once you know the dates of your exams you will know how much time is available to you. The best way to impress on yourself how much time you have available is to represent the time visually by using a diary, calendar, or simply a sheet of paper. In addition to the exam dates themselves, you need to include other fixed points, such as any remaining coursework deadlines, the final lecture, and any non-course commitments you have, such as holidays. All of this needs to be represented on a planner, such as the one shown in Figure 6.1.

Although your exams are likely to be scheduled within a defined period, it's probable that there will be some gaps between them, so you can stagger the start of the revision for each

Figure 6.1 Planning your time

April						
		1	2	3	4	5
6	7	8	9	10	11	12
13	14	15	16 Final lecture	17	18	19
20	21	22 Essay due in	23	24	25 Away for weekend	26
27	28	29	30			

May						
				1	2	3
4	5	6	7	8	9	10
11	12	13	14 Exam 1	15	16 Exam 2	17
18	19	20 Exam 3	21 Exam 4	22	23	24
25	26	27	28	29	30	31

exam accordingly. Knowing when your exams are will help you to appreciate how much time you have before the exams. The next step is to decide how you are going to allocate the time you have.

Decide how you are going to allocate your time

Deciding how to allocate your time is an important element of getting yourself organised. The following guidelines will help.

Decide what you are going to revise and how much

To decide how you are going to allocate time for revision, you need to decide what you are going to revise and how much you are going to revise it. You will already have some idea of this, having found out what you will be examined on (see 'Find out what you will be examined on' and 'Check you have all the necessary material' in section 6.1.1), but now you need to make some definite decisions about how you will fit the revision of your material to the time you have available. There are some important considerations to bear in mind at this point:

- if you are going to know your material in sufficient depth you will need to revise subjects more than once;
- however, that doesn't mean you have to revise absolutely everything (remember, it is usually better to have revised most things in sufficient depth rather than everything at just a surface level);
- you might also decide that you need to revise more for some exams than for others— you don't necessarily have to allocate equal amounts of revision time for each exam.

A useful process is to make an overview of each subject, listing the topics in that subject, and the headings under each topic (you will already have most of this information if you have identified whether you have the necessary material). This creates an index of what needs to be revised so that you can divide the revision into easy-to-manage sections.

Make a revision timetable

The kind of plan represented in Figure 6.1 gives you a useful overview but it doesn't give you enough detail about how you need to allocate your time. You need to zoom in on this over-view to create monthly, weekly, and even daily plans of what you need to do.

Imagine your four exams were as follows:

1. Comparative animal physiology: 14 May
2. Human genetics: 16 May
3. Biochemistry of enzymes: 20 May
4. Animal and plant diversity: 21 May

According to your plan (Figure 6.1) this gives you four weeks between your final lecture (16 April) and your first exam (14 May). However, there is also your final essay to submit on 22 April, and—assuming you will need the time between the final lecture and the submission date to work on the essay—that leaves just over three weeks to revise before your first exam. The remaining three exams then fall within the following seven days. These weeks need planning carefully if you are to cover the necessary material in the time available; this is where the weekly and daily plans come in (Figure 6.2).

Try this—Planning your revision

Two months before your next set of examinations, prepare a detailed revision timetable. Make it realistic by taking into account other commitments that you might have, such as part-time work and socialising.

Figure 6.2 Monthly, weekly, and daily revision plans

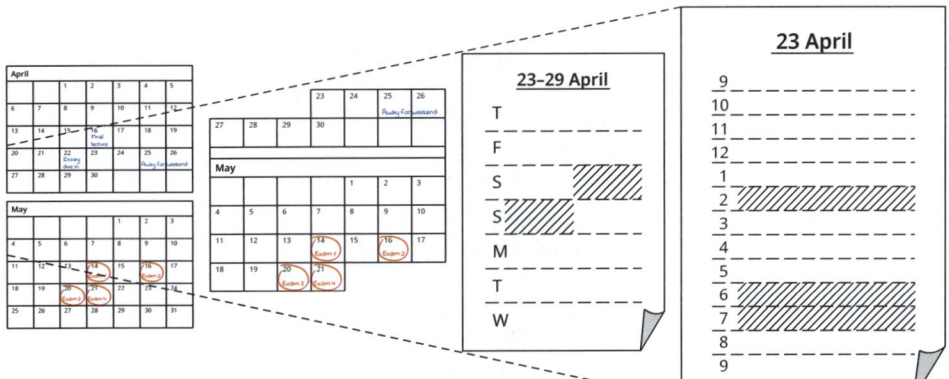

Be realistic and build in some slack

It is important to be realistic in your planning. Estimating clearly how much time you have should help you not to panic and so end up trying to cram too much revision into a short space of time. If you are clear how much time you have and what you need to do, you can then allocate your time realistically. A realistically planned revision timetable will have some slack built in so that when your plans need to change (for instance due to a certain topic taking longer than expected) there is space for this to happen. Of course, if the slack time is not needed you can use it for other things, for instance going over a topic again in more detail or perhaps rewarding yourself with an additional break. It's difficult to be precise about how much slack time is needed, but if you planned to revise seven hours in a day, you could leave the seventh hour unallocated to use for whatever was necessary. An example is shown in Figure 6.3.

In addition to planning in slack time you also need to plan in breaks: revising non-stop for seven hours is not an efficient use of time. The longer you go without a break, the more difficult it becomes to concentrate and absorb information. Breaks from revision, therefore, are not a luxury; they are a necessity if you are to be able to revise in a sustainable manner. Therefore, the morning and afternoon slots represented in Figure 6.3 shouldn't be three hours without a break, but rather shorter blocks with breaks built in. Keep your breaks short and free from unhelpful distractions; just getting up to walk round the room or make a drink is enough to renew your concentration. You also need to decide what times of day you work best; some people find working in the mornings easier, others find evenings more productive. Whatever your preference is, try to use it to your advantage, but don't use it as an excuse not to get up in the morning!

Try this—Planning your revision in detail

Using your overall revision plan as a guide, produce a detailed and realistic daily revision plan that you are confident you could stick to.

Figure 6.3 Daily plan with slack built in

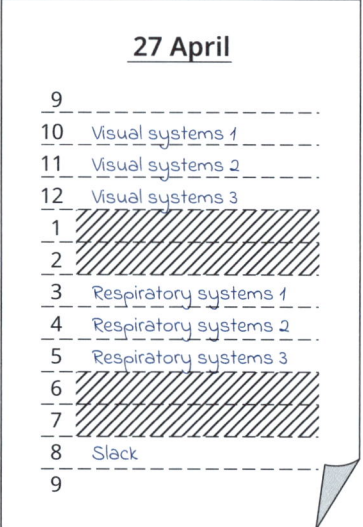

Do other stuff

Obviously it is important to be focused on your revision, but it is important to have a life outside of revision too. The longer your revision lasts, the more important it is to plan in time to do other things. Planning in time for other things isn't simply an end in itself but can be used to help you in your revision in a number of ways, for instance to:

• provide you with a break, so you can come back to your revision more mentally refreshed;

• re-energise you, so you can come back to your revision more physically refreshed;

• act as a reward for having achieved a certain target in your revision.

Precisely what you do to achieve these benefits will depend on what you find helpful. People often find exercise helpful, or doing something creative, or perhaps spending time socialising. You will need to be disciplined and not get carried away, but a short period of activity, perhaps each day, that you find enjoyable and would achieve some of the benefits listed will help keep your revision on track and sustain your motivation.

Use variety

In addition to spending short periods of time doing other things, another helpful strategy is simply to use a variety of approaches to your revision. You will know that this is helpful if you have ever tried to spend a revision session just reading, for example. Spending a large amount of time only using one approach can quickly become tedious and demotivating. But this isn't necessarily a sign that you need a break, it might just be a sign that you need to change your method. Try testing yourself with some questions or flash cards, or drawing a diagram, or writing a summary of what you've just read. We will deal with these and other suggestions in more detail shortly when we consider active revision techniques, but the important point to note here is the principle that using a variety of approaches to how you revise can be helpful. A change is often as good as a break (and better than a break if you are using breaks to avoid getting down to revision!)

Go public

Lastly in this section on allocating your time, consider making your revision timetable public; you could stick it up on a kitchen cupboard or on your bedroom door. There are a number of reasons this can be helpful:

- it reminds you of what your plans are;
- it informs others who share your accommodation what your plans are;
- it makes clear to others when you are available and when you're not;
- it makes you accountable to others ('I thought you were supposed to be revising?' or 'Shouldn't you be taking a break now?')

Whether or not you want to make your plan public will depend on who you live with, but it is definitely worth considering.

Check that your environment is suitable

Finally, in this section on getting yourself organised, you need to think about the environment in which you will revise. You might be the sort of person who always prefers to work in the same place every time, or perhaps you like variety. Some people find the campus library a helpful place to revise, whereas others prefer to study in their own room. Whatever your preferences, try to make decisions based on what will help your revision to be most effective. A suitable studying environment is one which is free from distractions. For example, it should be:

- well lit;
- quiet;
- not too hot or too cold;
- comfortable (but not to the extent that you fall asleep!);
- reasonably tidy and free from irrelevant clutter.

Sort these things out quickly before your revision session begins, but don't use them as an excuse to avoid starting!

Find out where to access support if you need it

There are a number of reasons why you might need to access additional support. It could be because you need help adapting to a new level of study, you have a disability or are neurodiverse, or you're experiencing exceptional circumstances. If any of these apply to you, your university's student support services are there to help.

6.1.2 **Use active revision techniques**

We said at the beginning of this chapter that you will know when your revision is being effective because it will make your material easier to understand, remember, and apply; this is what active forms of revision help you to do.

Active revision is any form of revision that makes you interact with the material in an involved and thoughtful way; this includes condensing your notes, drawing summary diagrams, testing yourself with questions or flash cards or gamification of revision using study apps, or online games and quizzes.

If you have ADHD, or are someone that struggles with focus, you might find that passive forms of revision, such as reading or copying, are more difficult to sustain for longer periods, and using active revision methods help to stay focused.

So, now you're organised, what are you actually going to do when you sit down for your first revision session? Assuming that you have planned your revision timetable so you have a reasonably clear idea of the topics you will be covering on any given day and how long your revision session for the day concerned will last, you are now in a position to begin to work on the detail. This will involve:

- filling in any gaps in your material;
- condensing your notes carefully;
- reviewing your notes regularly.

Fill in any gaps

The starting point, then, is to fill in any gaps in your material you have identified (as described in the section *Check you have all the necessary material*) and have decided what needs addressing. Your material could be supplemented from a number of sources: if you have simply missed a lecture, then borrowing a friend's lecture notes may be adequate or catching up with a lecture recording, but it is more likely that you will need to do some additional reading. This needs to be done in a certain sequence: make sure you understand the basics first before you progress onto more advanced issues. In practice, this means you need to start with lecture notes and handouts, followed by key chapters from core texts on your reading list. If you still need more detail then, once you are confident you understand the basics, you could move onto more specialist publications. This is where it is important that you have understood what you will be examined on (see section *Find out what you will be examined on*) so you are aware how much additional reading is necessary. There will always be more reading that you could do; the question you need to answer is whether there is more reading that you must do? At this stage, you need to be strategic in your approach: you won't have a lot of time to undertake detailed additional reading, therefore you need to identify exactly what is necessary and do that, but avoid doing more.

Try this—Identifying gaps in your notes

Find the learning outcomes or objectives of a recent course or module that you have taken. Now go through your notes and identify where in your notes each learning outcome or objective has been addressed. Identify which ones are missing from your notes and identify where you can find the missing information.

Condense your notes carefully

Once you have filled in any gaps in your notes you will probably have a lot of material. Therefore, a vital stage in any revision strategy is to condense your notes into a format that is more manageable. There are two simple reasons for this:

- the process of condensing your notes helps you to learn the material;
- the end product of condensing your notes provides you with a summary of your material which you can easily review (also enabling you to learn your material).

We will address the review part shortly, but first we will deal with the condensing element. Condensing your notes is an example of an active form of revision: it forces you to make decisions about which parts of your material are the most important and will help you to check how much you understand. This process consolidates your knowledge and helps you to make connections between different aspects of your learning and identify underlying principles. But how do you actually condense your notes? Condensing notes involves the following stages (and is illustrated in Figure 6.4).

- Taking your original notes on a topic (from various sources) and making a condensed version, perhaps on several sheets of paper.

Figure 6.4 Condensing notes

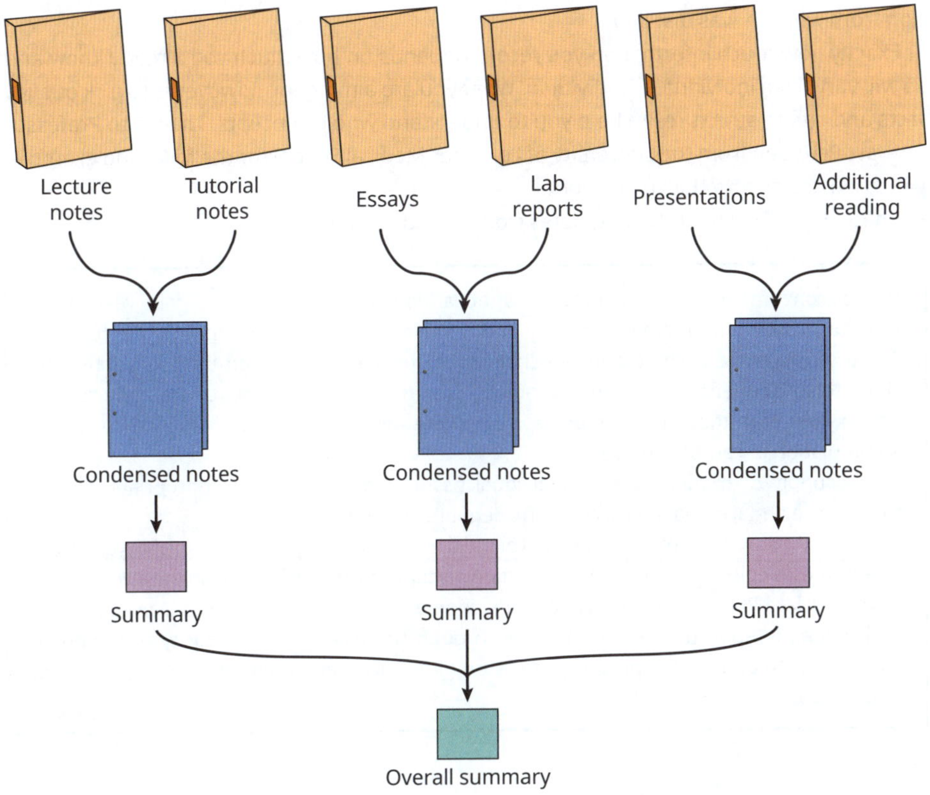

- Then taking this condensed version of your notes and trying to condense the information still further, perhaps to a single sheet of paper.

- Finally, writing an overall summary of the particular topic you are revising, perhaps on an index card.

Note, you don't have to condense your notes on paper. Using an online alternative (such as Microsoft Word, Microsoft PowerPoint, or Microsoft OneNote, Google Docs, Google Slides, or Google Keep) is also an option, but often the action and flexibility of writing things down is helpful, although, admittedly, it's not as accessible (e.g. via your phone when you're sitting on the bus) as online. So choose whatever suits you best.

The principle of condensing is simple enough, but how do you decide what information is sufficiently important for you to record and what you can leave out? Clearly you don't want to simply copy everything out (this would be a passive form of revision which would quickly become ineffective) but, equally, you don't want to be so sketchy with your notes that when you come back to them at a later stage there is insufficient detail for you to work from. This is why this process involves a number of stages: you can't skip from your lecture notes straight to an overall summary without going through the intervening stages. The reason for this is that, in producing condensed notes and then summaries of these condensed notes, you are not only producing a more condensed version of your material on paper but also structuring and connecting the information in your mind and making decisions about what is sufficiently important to record and what you can leave out. At each stage, you are attempting to distil out the essence of the particular topic you are studying to understand it better. What you will be left with, therefore, is not everything you need to know on the topic, but sufficient information to prompt you to recall the source information from which it was distilled.

Exactly how much information you record depends on how much you already know and so will vary from individual to individual. What you are aiming for, however, is to pick out key facts and create structures to help you to understand whole concepts. Thanks to Professor Stewart Petersen from the University of Leicester Medical School for the following example, which illustrates how this might work.

Imagine you had the following notes from a lecture on hypoxia.

Tissues consume oxygen and produce CO_2 at variable rates. The circulatory and respiratory systems must work together to supply and remove gases at an appropriate rate for each tissue of the body. Each step in the chain of supply can be affected by disease, but the ultimate consequence of any such condition is that some or all parts of the body receive less oxygen than they need—**hypoxia**, sometimes also associated with the inappropriate accumulation of carbon dioxide.

The supply of oxygen to each individual tissue depends first on appropriate oxygen content in arterial blood, and second on adequate local perfusion.

Appropriate oxygen content in arterial blood depends on effective diffusion of oxygen from alveoli into alveoli capillaries, and appropriate partial pressures of oxygen in the alveolar gas.

Appropriate PO_2 in alveolar gas depends upon adequate ventilation of the alveoli, which itself depends upon the ease of air flow through the airways, the capacity of breathing movements to generate appropriate pressure differences, and appropriate composition of inspired gas.

(Continued)

> Any disease process affecting oxygen supply to tissues must affect one or more of these steps, which therefore provides a convenient classification for pathophysiology, and the ultimate outcomes, local or generalised, are poor supply of oxygen—hypoxia.

Figure 6.5 Annotated notes

Hypoxia

Tissues consume oxygen and produce CO_2 at variable rates. The circulatory and respiratory systems must work together to supply and remove gases at an appropriate rate for each tissue of the body. Each step in the chain of supply can be affected by disease, but the ultimate consequence of any such condition is that some or all parts of the body receive less oxygen than they need – **hypoxia**, sometimes also associated with the inappropriate accumulation of carbon dioxide.

The supply of oxygen to each individual tissue depends first on appropriate oxygen content in arterial blood, and second on adequate local perfusion. Appropriate oxygen content in arterial blood depends on effective diffusion of oxygen from alveoli into alveoli capillaries, and appropriate partial pressures of oxygen in the alveolar gas.

Appropriate PO_2 in alveolar gas depends upon adequate ventilation of the alveoli, which itself depends upon the ease of air flow through the airways, the capacity of breathing movements to generate appropriate pressure differences and appropriate composition of inspired gas.

Any disease process affecting oxygen supply to tissues must affect one or more of these steps, which therefore provides a convenient classification for pathophysiology, and the ultimate outcomes, local or generalized, are poor supply of oxygen—hypoxia.

A useful way to start to summarise such notes is to annotate them as you read them, as shown in Figure 6.5.

You could also choose to summarise the content in your own words, as shown in Figure 6.6.

To condense the information further you could create a summary diagram, as shown in Figure 6.7. This summary diagram has the added advantage of including additional information.

In addition to considering the content of the information you are condensing, you also need to consider its format. The example used in Figure 6.7 is a deliberately visual example. This is because many people find it easier to remember something that is visual rather than something that is essentially text-based. In the lectures section of Chapter 3, we suggested a number of points to improve your note-making; many of these are also relevant for note-making for revision. Remember to try to use:

- headings and subheadings;
- colour;
- space;
- figures, including labelled diagrams and flow charts.

Try this—Producing a summary diagram

Think about the last lecture that you attended. Look at the lecture notes or handout for that lecture and summarise them into a single page summary. Now produce a summary diagram based on that lecture.

Figure 6.6 Summary notes

- tissues consume O_2 and CO_2 @ variable rates (circ. + resp. systems together supply + remove gases @ apt. rate)
- each step of supply can be affected by disease → less O_2 than needed (hypoxia)
- supply of O_2 to tissues depends on
  1. apt. O_2 content of arterial blood
  2. adequate local perfusion
- 1 depends on effective diffusion of O_2 from alveoli → alveoli capillaries & apt. PO_2 of O_2
- which depends on adequate ventilation of alveoli
- which depends on ease of air flow thro' airways, breathing movements + comp. of insp. gas

NB. diseases affecting O_2 supply to tissues must affect at least 1 step

Figure 6.7 Summary diagram

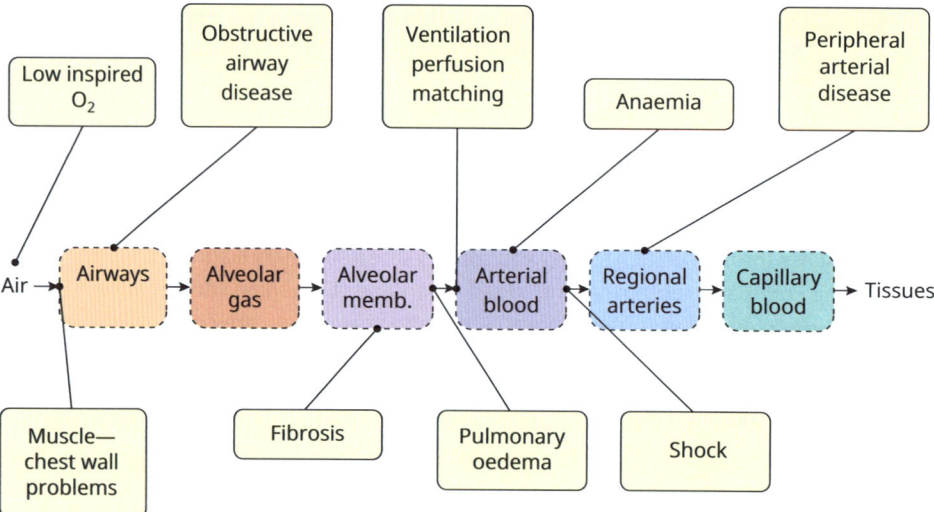

Review your notes regularly

We observed earlier that condensing your notes provides you with a summary that you can easily review. If you have ever tried to review uncondensed notes you will know that it is a very difficult exercise to sustain because of the sheer volume of information you are trying to read. Even if you manage to review uncondensed notes for a particular topic, the more topics you have, the more unsustainable it becomes. If you have produced summaries, however, not only do you have your notes in a handy format for you to review, as identified in *Condense your notes carefully*, but the process of condensing your material will also have helped you to learn it. Clearly you won't remember everything (that is why the review stage is important) but you will remember some and you will also have something that is reviewable.

The importance of reviewing your notes regularly can be highlighted by considering what happens when you don't do so. What inevitably happens is that the longer you go without reviewing your material the more you forget. Without regular reviews of what has been revised you won't retain the information you have taken the trouble to learn.

The relationship between reviewing and recalling information is straightforward enough, but how do you actually review your material? There are many different methods of reviewing information but, whichever method you choose, it needs to test your recall of a topic by:

- checking your ability to understand central concepts; or
- remembering key facts; or
- linking the information to other topic areas;
- or perhaps all three!

As we have already noted, you won't remember everything, and you can always go back to your original notes to remind yourself of particular details; the important thing is to do the reviewing. Methods of review include:

- testing your recall of key facts with index cards (or flash cards), putting to one side the cards you can remember and repeating the cards that you can't;
- reproducing from memory headings and keywords for a learnt topic area and then checking them against your notes;
- recording yourself explaining concepts or theories to an imagined audience, then playing back your explanation, checking it against your notes;
- making a visual summary (like the flow chart in Figure 6.7) to summarise information you have remembered;
- reproducing important diagrams that summarise large amounts of information or whole concepts, for example metabolic pathways;
- enlisting the help of a friend to test your recall using any of these methods;
- putting up summaries of key information around your home to give you the opportunity for on-the-spot tests;
- testing yourself with question banks and past papers (more on this shortly).

As you can see, there are many options; choose whatever suits you best.

Reviewing your material does not need to take a lot of time. A good time to review material is at the beginning of a revision session but it can also be fitted into any spare time during the day. Frequently reviewing your material will help to build up your confidence and reassure you that you are making progress.

6.1.3 See the big picture

We have noted, in section 6.1.2, that condensing your notes supports the consolidation of your knowledge and helps you to make connections between different aspects of your learning and to identify underlying principles. However, when paying attention to detail, it is easy to lose sight of the context and principles. It is important, therefore, that you make a special effort to appreciate the big picture. Try not to get so focused on the details that you fail to see either where the information fits in to the larger context or perhaps a simple principle that underpins it, both of which would make it easier to understand and remember.

The exact parameters of the big picture will depend on how much breadth and depth you are required to know, and this will change depending on the module choices you make throughout your course and the level of study you are at. However, context and principles are important to appreciate and identify because they help you to develop a means of structuring, storing, and retrieving information. They also help because exam questions rarely ask the precise question you had been hoping for; rather they require you to apply your knowledge to different situations.

Imagine looking for something you saved somewhere; maybe in a file in Microsoft 365, maybe in a Google Drive file, or perhaps in a note-making app on your phone, or one of the many browser windows you have open at any given time. When you have multiple places that you can create and store things in it can get very difficult to remember or find where something specific is located. If you did manage eventually to find the correct piece of information it would be difficult to then explain how the information relates to all the other information you have stored in different places.

Now imagine the same information but this time organised appropriately a single online space, with appropriate folders and sub-folders. Suddenly it becomes much easier to not only find the required information, but also to relate it to other information. Seeing the big picture is a bit like storing files into an organised file store; it enables you to appreciate how information fits into a wider context and to identify the principles that underpin it. Let me know.

6.1.4 Practise outputting the information you have learnt

We have noted the importance of understanding, remembering, and applying the information you learn. However, the steps we have focused on so far are largely concerned with the first two elements: understanding and remembering. If you are going to perform well in the exam room (more on this in section 6.3 *Sitting exams*) then you need to be able not only to understand and remember information but, importantly, you need also to be able to apply it. This is because exam questions, whatever the format (and this includes 'open book' exams), are specific and you are therefore required to give specific answers. When you open an exam paper you're very unlikely to find a question which reads 'Please tell us everything you know about comparative animal physiology' (although it's surprising how many answers are written as if this was the kind of question asked!) A much more likely question would be 'Compare and contrast the retinal organisation of a named mammal and reptile'. Therefore, the information you have taken time to understand and remember needs to be applied to the specific question. This particularly applies to open book exams because the lecturer has set the question assuming you can look up the topic, so the grading will be weighted towards your ability to apply what you have learned to answer the question (and not the memorisation of the subject). Spending the time revising, condensing, and organising the content is also essential, even if books are available, as it takes longer to look things up than to recall them. So, the learning process remains very important regardless of the format of the exam.

It would be very unfortunate if you had taken the time and effort to understand and remember information but failed at this last crucial step. The best way to improve your ability to answer questions specifically is to practise. The two main ways of practising are:

- creating question banks;
- using past papers.

A note on 'Seen Question Exams'

It is also worth noting here that lecturers are increasingly setting 'seen' exams, where the questions are released prior to the exam and students are given time to prepare an answer. For example, students might be given seven potential topics, of which three appear in the actual exam (or whatever combination of numbers). It is imperative that you prepare and practise outputting for these exams, as the expectations of what you will be able to create are much higher than under unseen conditions. It might help to think of Seen Exams more like writing a coursework essay and use the guidance in later sections for these. Try to ensure that you have developed and memorised solid plans for your seen questions before entering the exam room.

Create question banks

A question bank is simply a collection of questions that you think you might be examined on. The questions shouldn't be full essay questions, but short, one-line questions that test very specific knowledge. Question banks are useful for a number of reasons:

- past papers are not always available to you;
- the format of the exam might have changed;
- thinking up questions helps you to reflect on how much you know;
- answering the questions helps you to apply your knowledge.

Question banks can be created from any questions that you think you might need to know the answer to. A good place to start is the learning outcomes for the modules: use these as a prompt to help you think of questions. Keep adding questions to your question bank throughout your course; questions prompted from lectures, or tutorials, or reading.

You can then use these questions to check your knowledge of the material and so monitor your progress, or as a means of review, or to supplement your use of past papers. It is also beneficial to team up with some fellow students to share the task of creating a question bank; this will save you time as individuals and give you a greater variety of questions than you might have created on your own.

Try this—Creating a question bank

Get together a small group of your fellow students on a course you are revising for. Ask each of them to come up with three questions that they think might be important to be able to answer based on the learning outcomes or objectives of the course or module. Share these questions between you to form a substantial question bank.

Use past papers

Past papers are an extremely helpful resource when you come to practise outputting the information you have learnt. If you have already found out what you will be examined on (see *Find out what you will be examined on* in section 6.1.1) you will know what the format and scope of the exam will be. Assuming there are no major changes to an exam's format or scope for that particular year, you can assume that the past papers will be representative and so be a good test of your learning so far. There are a number of ways you can use past papers, including:

- practise analysing questions and setting up plans to answers to test your recall and ability to adapt material to the set question;
- practise writing full answers, by hand (i.e. not typed), in the time allowed—as most exams are still handwritten this will also help you practise writing legibly over an extended period and under time pressure;
- review your answers by checking them against your notes and highlighting any missed or inaccurate information;
- ask one of your lecturers to read your essay and comment on it;
- analyse your answer to see how it could have been improved. (Did it need more information? Did it have a logical structure? Was your expression clear? Did you stick to the question?)

Your precise use of past papers will depend, in part, on the format of the exam: multiple choice, short answer, essay, or open book. We will deal with this in more detail in the next chapter.

6.2 Just before exams

If you have got yourself organised and managed to do a reasonable job of your revision, as described earlier, then there will be less of a temptation to do any last-minute cramming. You need to be realistic at this point though: there will always be more you could do, but at some point you need to stop.

It is very unusual, even for students who have managed to get themselves very organised, to be able to walk into an exam room feeling completely prepared. After all, exams are—by their very nature—an unknown quantity (even 'open book' exams have an unknown element to them). It's therefore common to want to try to cram in a lot of revision just before an exam (either because your revision hasn't gone as well as you had hoped or because you're anxious), but if you do find yourself in this situation then be careful; there is work that is helpful to do at the last minute, but, equally, some which can be unhelpful.

6.2.1 The wrong sort of last-minute work

Last-minute cramming is the wrong sort of last-minute work. There are a number of reasons for this.

Last-minute cramming is ineffective

If there is material that you still haven't covered the night before the exam you are unlikely to gain much benefit by covering it at this late stage. Your time would be much better spent going over things that you are already familiar with (see section 6.2.2).

Last-minute cramming makes you forget other stuff

Last-minute cramming is often counterproductive because trying to cram in a lot of information (particularly new information) at the last minute can cloud your memory of what you already know and so has the effect of pushing out other information that you may have otherwise been able to retain.

Last-minute cramming makes you tired

Exams are hard work and require a lot of energy and concentration. Therefore, making yourself tired by staying up late the night before (or worse, through the night) is not good preparation. Your time would be much better spent trying to get some rest so that you are alert for the exam and so can make a good attempt at answering the questions, rather than being so tired you can't think clearly.

Last-minute cramming makes you worry

Last-minute cramming tends to make you focus on what you don't know rather than what you do know. Thinking about all the things you don't know the night before an exam is a bad idea! It makes you worry, it gets you stressed, it makes it difficult for you to get to sleep, which makes you more worried and more stressed . . . and so it goes on.

6.2.2 The right sort of last-minute work

While last-minute cramming is the wrong sort of last-minute work, there is work you can do just before an exam which is helpful.

Plan your use of time

It is important to plan how you will use the time you have in the exam before you get to the exam room. This doesn't have to be a last-minute exercise but, if you haven't done it as part of your revision strategy, now is the time to do it. You need several pieces of information to plan your use of time effectively:

- how long the exam lasts;
- how many questions the paper contains;
- how many of the questions you are required to answer;
- whether all the questions are worth the same number of marks.

Once you know this information you can decide how much time you will spend on each question, remembering to allocate time for choosing questions (if you have a choice), analysing questions, planning answers, and reviewing answers, as well as (of course) time to write your answers. For example, if you had a one-hour short answer paper containing nine questions, each of which was worth 10 marks, you might work out how to allocate your use of time as follows:

- 1 hour = 60 minutes;
- 5 minutes per question (including analysing and planning time) = 45 minutes;
- which leaves 15 minutes checking time;
- 15 minutes checking for 9 questions = just over 1.5 minutes checking time per question.

If you had a three-hour essay paper containing 10 questions, all worth 25 marks, of which you had to answer three, you might allocate your use of time as shown in Figure 6.8.

Planning your use of time before the exam will give you one less thing to think about in the exam itself. It will also ensure that you at least attempt all the questions you are required to attempt, rather than simply focusing on the ones you are confident on. In an essay paper where you are required to write three essays, for example, you will almost always get more marks by having a go at all three essays rather than spending time trying to make two of them really good ones. Even if you are not confident about the third essay and think you can only write a fairly poor answer, the chances are you will get more marks for attempting an answer to the third one rather than by refining an existing one.

The night before the exam

We have already noted that the night before an exam is not a good time for learning new information. It is, however, a good time for practising your exam technique or checking your knowledge of what you have already learnt. So use your question banks and past papers to go over the information you have already covered, making sure that if you come across something that you don't know you look it up so you do know it. Going over pre-revised material in this way the night before an exam will both boost your confidence regarding how much you do know (you might be surprised!) and act as a reminder of the things that you had learnt but couldn't recall in response to a question.

Figure 6.8 Example plan for use of time for a 3-hour essay paper containing 10 questions of which you are required to answer three questions

Time (minutes)	Task
5	Read questions
5	Choose questions
50	Question 1 • analysing (5) • planning (5) • writing (30) • reviewing (10)
50	Question 2 • analysing (5) • planning (5) • writing (30) • reviewing (10)
50	Question 3 • analysing (5) • planning (5) • writing (30) • reviewing (10)
20	Checking

Do make sure, though, that you don't work late on the night before the exam and try to leave some time for relaxing before you go to bed as a good night's sleep is important for doing your best the next day.

The morning of the exam

The morning of the exam is also not a good time for learning new information. As we have already mentioned, cramming can make you forget other stuff; in particular it can cloud your ability to remember overall concepts. Instead, focus on reviewing main points to give you an overview of what you have learnt. If your exam is in the morning, it would be a good idea not to do any sort of revision on the morning of the exam at all.

6.2.3 Other last-minute preparations

Good exam technique is about more than simply learning information and then outputting that information in response to specific questions. Your preparation could have gone brilliantly, but if you are so tired you can't think clearly, or you turn up at the wrong location, your preparations will have been wasted. So here are a few more last-minute preparations to consider.

Get your stuff together

Getting everything you will need for the exam organised the night before will give you one less thing to think about on the day itself. You will need a pen you can write with comfortably for a long time, and at least one spare. Depending on the exam you may also need other equipment, such as a calculator. Also think about what you don't need, which is anything that might cause an invigilator to suspect you of cheating if they found it on you. This includes mobile phones, audio players, scraps of paper with notes on, and scribbles on the back of your hand. It is particularly important for open exams that you know exactly what kinds of material you are expected to bring in with you; for example, this might be a single folder containing your lecture notes or it could be annotated scientific papers that you need to explain or any number of other materials. Either way, just make sure you know what the rules are.

Try this—Preparing for an exam

Several weeks before your next exam, find out the date, the time, and venue. Find out how long before the exam starts you will be allowed into the exam hall and what equipment you can take in with you. Make sure you know how many questions are on the paper, how many questions you have to answer, and how long you have to answer each one.

Double check when and where you need to be

Another thing you don't want to be worrying about on the morning of the exam is when and where the exam will take place. Make sure you have the latest information on the date, time, and location of your exam (sometimes these change due to timetable clashes, which is why the latest information is important). Then double check this with other people who are taking the same exam, just to make sure. If your exam is taking place in a location you are not familiar with, it is worth either going to the venue beforehand or arranging to go with someone on the day who is sure they know where it is. Also, set your alarm to make sure you wake up in time, and then set another alarm, just in case. You may also want to let the people you live with know when you need to be up, so they can help make sure you're up too. Having more than one wake-up call reduces the chances of you worrying about not waking up in time and so makes it more likely that you will get to sleep. Plan to arrive at the exam room in plenty of time, as racing in at the last minute (or late) flustered and out of breath is not a good way to start.

Get some sleep

We have already identified that last-minute cramming isn't helpful, but a good night's rest the night before the exam is. However, this is easier said than done, so it helps to have a strategy. Firstly, make sure that you don't drink too much caffeine too late, as clearly this will keep you awake. Secondly, it is very unlikely that if you stop working at 11pm and then roll straight into bed you will fall asleep quickly, and the longer it takes you to fall asleep the more worried you will become about not being able to sleep and so the more awake you will be.

To help you relax and sleep well, plan something relaxing to do in between stopping work and going to bed. It could be chatting to your housemates, going for a walk round the block, or watching a bit of television (as long as you don't start a whole film!) Anything that will help stop your mind racing and calm you down a little will make it more likely that you will be able to sleep.

Manage your stress levels

We have deliberately given this section the title 'Manage your stress levels', rather than 'Get rid of stress' because, firstly, it's unlikely that you will be able to get rid of stress altogether, and, secondly, some stress is helpful: it is your body's way of preparing itself to cope, making you more alert and attentive. All that we have covered so far in this chapter, and the previous chapter on revision skills, will help you to manage your stress levels, because one of the most stressful things, as far as exams are concerned, is lack of preparation. So getting organised will really help: the most stressful exams are always the ones you haven't got organised for. If you still feel anxious though, it can also be useful to talk to someone about your worries concerning exams. This could be a friend, your tutor, or someone in your university's student support services. Additionally, on the day of the exam itself, there are a number of things that will help you manage your stress levels:

- try to avoid talking to other students if you think it will increase your anxiety levels;
- get plenty of fresh air—exam rooms are often stuffy places, so get plenty of fresh air beforehand (this will also help you avoid talking to other students);
- try some breathing exercises, by taking conscious control of your breathing (for instance, breathing in for four seconds and out for six seconds) can reduce stress levels and help combat anxiety;
- make sure you use the toilet before the exam—you really don't want to be needing to go halfway through the exam: it will be a major distraction, and will make you feel unsettled (and, therefore, stressed).

Exams are physically tiring. So there are also a couple of things not directly related to the management of stress that you should keep in mind to make yourself as prepared as possible from a physical point of view.

- Try to eat breakfast—not everyone finds this easy, but exams are hard work and you need energy. At the very least, have something simple but high-energy, like a banana.
- Drink water—to keep you hydrated and help you think clearly (and take a bottle of water in with you).

6.3 Sitting exams

Finally in this chapter, sitting exams; what are some of the things you should or shouldn't do when you get to the exam room, and how can you output effectively all that knowledge you have learnt?

6.3.1 Arriving in the exam room

When you arrive in the exam room, there are several things you can do to get yourself off to a good start.

Get settled

Exams have a lot of build-up and there is often a lot depending on them. So, it is inevitable that you will be feeling at least a little anxious. Just by sitting down, taking a few deep breaths, getting out your pens, your clock or watch, and your bottle of water, and generally getting yourself comfortable in your chair will help you settle down.

Read the instructions very carefully

Once you have been told by an invigilator that you can turn over the paper, turn it over. The first thing to check is that it is the paper you were expecting (if it isn't, put your hand up—often exam rooms contain several different exams at once; you could just be at the wrong desk). Once you are sure you have the right paper begin to read the instructions carefully. The instructions should be familiar to you from your preparation, but check in particular the number of questions, how many questions you are required to answer, and how many marks each question is worth.

Remind yourself of your time plan

You should have planned your use of time beforehand (see section 6.2.2), so this should be simply a case of reminding yourself what you had decided to do (but if you haven't—do it now). You may find it helpful to make a note of your time allocation on the question paper so you don't forget or become confused about what you had planned to do (which is easily done when you are under pressure).

6.3.2 Answering the questions

The areas we have covered so far in the sections on *Just before exams* and *Arriving in the exam room* apply equally to all types of written exam, whether multiple choice papers, short answer papers, or essay papers. However, when it comes to actually answering the questions, there are some exam techniques that are particular to different types of written exam. We will deal first with some general advice applicable to all question types and then address how these apply specifically to different types of exam paper.

General advice for all question types

Regardless of the type of exam paper you are faced with, it's a good idea to have a clear sequence of steps in mind to help you answer the questions to the best of your ability. We suggest the following:

- analyse the questions;
- plan your answers;

- write your answers;
- review your answers.

   Following these four steps will help ensure that you have a measured approach to what can otherwise be quite a panicky situation.

Multiple choice papers

Of the question types we have identified, multiple choice papers require the least amount of analysis and planning. However, that's not to say you can dispense with analysis and planning altogether—you just need to approach it in a slightly different, and briefer, sort of a way.

Analyse the questions

Usually, with multiple choice papers, you will not have a choice of questions you can answer. If this is the case then there is no need to read through the entire paper before you begin. Instead, simply read the instructions carefully (as described in *Read the instructions very carefully* in section 6.3.1) and make sure that the structure of the paper is what you expected. Once you have confirmed this, you can begin to work on the individual questions. With multiple choice papers you should normally start with the first question and then work through the questions in sequence. When you come across a question you don't know the answer to, don't dwell on it too long, just move on to the next one and come back to the ones you are not sure about at the end.

Plan your answers

You neither have, nor need, much time to plan your answers for multiple choice papers. However, a few moments spent planning will help make sure you don't rush headlong into a wrong answer. A good approach is as follows:

- read the question carefully;
- don't look at the possible alternatives yet—try to answer the question independently of the options given;
- look down the list of options to see if your answer is there;
- read all the other alternatives just to make sure.

Write your answers

For multiple choice questions this is simply a case of ringing a response or ticking a box, but do make sure that you have read the question carefully. Sometimes multiple choice papers include multiple response questions as well as multiple choice questions, that is to say questions that require more than one answer. For example:

Which of the following statements about the neuron is true?

a) The resting potential is dependent on the membrane permeability to Na^+ ions ☐

b) The resting potential lies close to the equilibrium potential for K^+ ions ☐

c) The membrane is only permeable to Cl^- ions ☐

d) The membrane is impermeable to K^+ ions ☐

In this question, as indicated by the stem: 'Which . . . is true?', there is only one correct answer, (b).

In some questions, there may be more than one true statement. These are indicated in two ways. For example:

> Which of the following statements about the lungs are true (tick all that apply)? . . .

Or some take a more complicated form as follows:

> Which of the following statements about the ventilation of the lungs are true?
>
> a) The expansion of the lungs is mainly as a result of contraction of the diaphragm.
>
> b) The expansion of the lungs has to overcome the resistance due to surfactant.
>
> c) Expiration at rest is the result of contraction of the intercostal muscles.
>
> d) The tidal volume is typically 500 ml.
>
> (a), (b), and (c) are true   ☐
>
> (a), (b), and (d) are true   ☐
>
> (b) and (c) are true   ☐
>
> none of the above   ☐

This second type of question is quite complicated and you need to work through the statements carefully, marking each one that you think is true and then seeing how your list matches with the options given.

Also, make sure you have understood the sense of the question. For example, check whether the question is asking you to indicate which is true or which is not true.

Your exam questions may also be set out as Extended Matching Sets (EMS). A simple version of an EMS question is:

> 1. During quiet respiration, the main contributor to inspiration is contraction of the .... During forced expiration the passive relaxation is supplemented by contraction of the .... and the .... muscles.
>
> 2. In actively metabolising tissues, the pH is .... and the oxygen dissociation curve shifts to the right, leading to .... affinity. This is termed the ....
>
> 3. During heavy exercise, the arterial pCO_2 may be .... as a result of the ventilatory response to the production of ....
>
> 4. Taking several deep breaths prior to diving extends underwater endurance because the .... of the blood is initially......................
>
> (Continued)

External intercostal	Decreased
Pectoral	pH
Bicarbonate	Lactic acid
Increased	Diaphragm
pO_2	Internal intercostal
Bohr shift	Chloride shift
Abdominal	Tidal volume
Unaltered	Buffered

Here you have been given a number of incomplete statements and a table of words that could fit those statements. You have to complete the statement with the most appropriate word, bearing in mind that each word may be used once, more than once, or not at all. Again, the best strategy is to read through the statement and think what should go in the space first, then look at the list and see if any of the words match. That way you are less likely to be confused by the distracters.

Review your answers

When you get to the end of the paper, check back through the paper and re-read those questions that you were not originally sure about, making sure that you have answered all the questions. The only time when you should change this strategy is if you have a paper where the questions have negative marking (this type of paper is not very common): in this case, incorrect answers will actually cost you marks, so you may be better off leaving some questions blank if you are really unsure of the answer. After having checked for uncompleted questions, review your answers, starting with the ones you were least sure about. If you amend an answer make your change very clear; there will be advice on how to do this in the instructions at the beginning of the paper.

Short answer papers

Short answer papers can also be tackled using the same four steps. Again, one of the keys to doing well on these papers is being strict with your timing—don't spend a lot of time thinking about each question. If you know the answer, answer it quickly, if you don't know the answer, move on and come back at the end, when you may have time to think about it more carefully.

Analyse the questions

As with multiple choice papers, short answer papers often don't give you a choice of which questions you can answer. Again, if this is the case, you don't have to read through the entire paper before you begin. Start with the first question and read it carefully to make sure you understand it, perhaps underlining key words to help you focus on the exact meaning. If, after a quick reading, you are not sure how to answer it, move on to the next question and come back to the ones you are less sure of at the end.

Short answer questions usually require specific factual information for the answer, and not detailed description. Usually the number of marks to be awarded is given on the exam paper, so that can be a guide to the amount of information you need to give. Typical short answer questions might take the following forms:

1. Draw a labelled diagram of the nerve action potential. (4 marks)
2. Label structures A–D on the diagram of the ear. (4 marks)
3. List the stages involved in the replication of DNA. (5 marks)
4. Describe the process of synaptic transmission. (5 marks)

Plan your answers

With short answer questions, in briefly planning your answers you are trying to make sure that you select information relevant to the question and put it into a logical and coherent order. In each of the questions in the box, the marker is looking for very specific factual information. To answer the question, you will need to note down a few keywords to stimulate your memory and organise your thoughts.

Write your answers

Use your brief plan to prompt you to make your points and structure your answer. You only have a short time to write each answer, so the important point is to make sure that all the key facts are included. For example, the diagram of the nerve action potential does not need to be a great work of art, but you must remember to include key details such as scales to mark the voltage and the time as well as the actual factual content. When writing your answers, brevity and factual content are the keys to a good answer. It is also important that you write legibly—this can be difficult if you are in a hurry and if you are not used to handwriting for long periods (which is why we said in the previous chapter that it is important to practise writing by hand for extended periods). Take some time to make sure your handwriting is legible.

Review your answers

Check that you have answered every question that you needed to answer (probably all of them), then review your answers. As with multiple choice papers, start with the questions you are least sure about because these are probably the ones you will be able to improve the most.

Essay papers

Often with essay papers you will have a choice as to which questions to answer. If this is the case then make sure you read through all the questions carefully before you choose which one (or, more likely, ones) you are going to answer. As you read the questions, mark the ones that you think you will be able to answer the best, then go back and read them again to make a final choice. As with multiple choice and short answer questions, start with the questions you are most confident with to get you off to a good start and stimulate your thinking.

Analyse the questions

Analysing questions is particularly important for essay questions. With multiple choice and short answer papers your answers will usually just be facts or isolated pieces of information. With essay questions, however, there is more work to do to at the analysis stage to organise your answers logically and coherently. Essay questions will usually be very specific, so, instead of asking 'Please tell us everything you know about comparative animal physiology', they are much more likely to ask you to, for instance, 'Compare and contrast the retinal organisation of a named mammal and reptile'. Applying your knowledge to this particular question requires you to analyse the question briefly, using the same technique that we will identify in Chapter 7, section 7.3 *Analysing the question*. The key information to identify in the question or title is:

- the subject of the question;
- the instruction;
- the key aspect;
- other significant words.

Plan your answers

Planning answers is particularly important for essay questions because it helps you to:

- select information that is relevant to the question;
- put this information into an order that is logical and coherent;
- write your ideas down at an early stage to help you remember your key points;
- monitor how much information you are covering in the time allowed.

Write your answers

When answering essays under exam conditions it is important that you keep to your time plan, as it can be easy to lose track of time if you get into the flow of writing. One way to do this is to make a point of checking the time when you come to the end of a section in your answer. This is preferable to checking the time at random intervals (which could be insufficient) or too frequently (which could distract you from writing fluently). Also, keep looking at the question to remind yourself of the focus of your answer, and keep looking at your plan to remind yourself of what you need to cover. Again, the end of each section of your answer makes a good natural break for checking that you're still on track. Make sure that as you write, particularly as you get into the flow of writing, that you keep your handwriting legible.

Review your answers

Being able to review your answers is where the benefits of having planned and monitored your use of time become obvious. Not only should you have had enough time to make a reasonable attempt at each of your chosen questions, you should also have time left at the end to go back and review your responses. Making time to review your work can make a significant difference to your final grade.

As with multiple choice and short answer questions, first check that you have answered every question that you need to answer, then review your answers, starting with the one you were least sure about. The purpose of reviewing answers in an exam context is slightly different to reviewing answers in a coursework setting: you don't need to finish up with a very polished piece of writing which is spelling error-free and grammatically faultless, just one that answers the question well and is legible. Of course, if you do end up with a polished piece of writing then that's good, but it's not the primary purpose of a review in an exam context.

Instead, you should be looking at the bigger issues, such as:

- Have you answered the question?
- Does the order of your material make sense?
- Is the meaning of what you've written clear?
- Have you made appropriate use of figures to supplement what you've written?

These are the main issues; you will get more marks for a response that has spelling and grammatical errors but answers the question well than you will for a response that has perfect spelling and grammar but doesn't answer the question well.

If you need to make significant structural changes to your answer, such as moving a paragraph from one section to another, you need to do this in a way that does not confuse the people marking it. In a word-processed script this would be straightforward, but in a hand-written script you will need to annotate it in a way that makes it clear you are moving a section rather than, for instance, deleting it.

 ## Chapter summary

We said at the beginning of this chapter that effective revision makes the material you are revising easier to understand, remember, and apply. Getting yourself organised in your revision, with a clear timeline leading up to the exam, is a crucial first step in revising effectively and can reduce the amount of stress that exams cause. Making your revision an active, rather than passive, process will make your revision both more interesting and more effective. Making sure that you don't lose sight of the big picture in terms of how detail fits into an overall structure is also important, as is practising outputting the information you have been learning to ensure you can apply it to specific exam questions.

Performing well in exams requires particular skills. Although your exam performance will be heavily influenced by how well you have revised, it is how you conduct yourself during an exam that will ultimately determine your grade. What you do just before an exam and when you arrive in the exam room are important stages of your preparation. Different types of question require slightly different techniques, but all require you to analyse questions, plan answers, write answers, and review answers.

7 Preparing for coursework assignments

Introduction

Almost all programmes of study include coursework assignments. As we discussed in Chapter 4, preparation for assessments will make up a significant proportion of your independent study time and they are important because it is through successful completion of the coursework assignments that you demonstrate your achievement of some or all of the learning outcomes for your course.

Coursework assignments are designed to enable you to develop your learning and understanding of the topic, to practise different forms of presentation, and also to provide guidance as to how you are meeting the learning outcomes for the course.

In this chapter, we will look at the different types of assignment you may encounter during your studies and the aims of these assignments. We will then explore how to analyse the questions you are presented with and consider some of the common problems you may face when tackling your coursework, and how to avoid them.

7.1 Types of assignment

Your coursework assignments may be *summative* or *formative*:

- Summative assessments: you should receive constructive feedback on your work and the marks awarded will contribute to your overall mark for the course.
- Formative assessment: won't contribute any marks for your performance but you should be provided with constructive feedback to help you develop in preparation for summative assessments.

Note that both summative and formative assessment include the provision of feedback by your lecturers (see Chapter 13, section 13.2 *When do you get feedback?*) so that your learning includes both the process of undertaking the work but also important feedback on what you did well and how to improve in the future.

Assignments come in a wide variety of forms, including:

- essays;
- reports;
- oral presentations;
- poster presentations;
- multiple-choice question exercises;

- data analyses;
- dissertations;
- literature reviews;
- critical analyses.

This list is by no means exhaustive and some of the activities listed may form a part of larger activities, for example, you are likely to need to analyse data as part of a report of laboratory practical work, or a fieldwork exercise. Likewise, a review of the literature is likely to be required as a part of most of the other assessments, in particular dissertations.

Whatever the form of the assignment, there are stages that are common to preparing all assignments, which we will look at in general terms before considering how to approach specific types of assignment. It's vital that you refer to feedback on previous work; this is something that you should bear in mind throughout the various stages, which include:

- analysing the brief;
- researching the topic;
- making a plan;
- writing a first draft;
- reviewing and redrafting;
- proofreading.

In the next sections, we will consider the broad aims of assignments and how to go about analysing the question before thinking about the common problems, many of which are linked to the planning process. Then, in the following chapters, we will look at the processes of planning in more detail and drafting and researching your assignment.

7.2 Different assignment aims

As we will explore in more detail below, assignments are presented with specific briefs that set out what is expected of you and it is important to analyse that brief carefully to make sure your work meets the expectations.

In principle, there are two main aims for coursework assignments:

- to discuss a topic based on a combination of course materials and independent investigation of the research literature. Most often the topic will be presented as a question that needs to be answered. Examples of such assignments include essays, dissertations, and oral presentations.
- to collect your own data, present and analyse the results, and interpret them in the context of the research literature. Examples of such assignments would be laboratory, fieldwork, or project reports.

Increasingly, the assessment regime for a programme is designed around the use of 'authentic assessments'. These are assessments that are designed to reflect the way in which learning may be used in real life. For example, rather than writing an essay about a topic, you may be asked to prepare a report for members of the public concerning a current high profile issue in the biosciences. These types of assessment are seen as valuable because they involve demonstration of a range of skills, that will be relevant to professional careers, alongside your understanding of the subject matter. However, authentic assessment will usually require some combination of researching a topic and working with data.

7.3 Analysing the question

It may seem obvious, but it is very important that you answer the question that is asked of you: there is no point in preparing an assignment that doesn't address the question asked, however brilliant that piece of work may be. So, you have your time plan in front of you, and are ready to go. STOP. Before you do anything else, read the question carefully and then read it again, highlighting the following features:

- assignment format;
- the subject of the question or title;
- the instruction;
- the key aspect;
- other significant words.

Figure 7.1 and Figure 7.2 show a couple of worked examples for you.

Sometimes, it might seem that you are having to highlight almost every word, but it is still worth doing if it helps you register all the important pointers.

Figure 7.1 Analysing the question or title: example 1

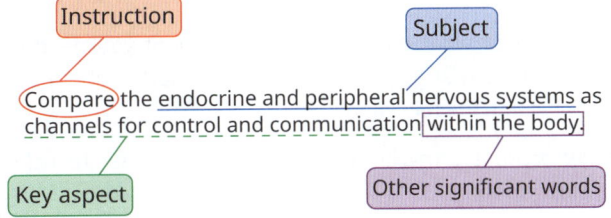

Figure 7.2 Analysing the question or title: example 2

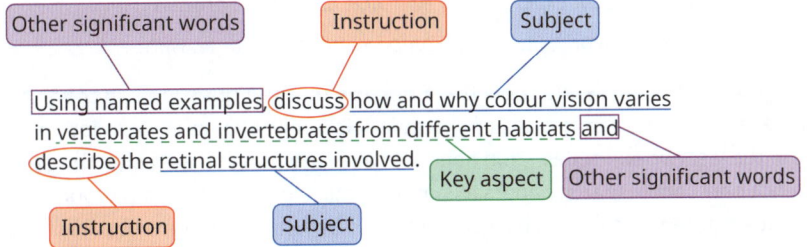

There are several standard instructions used for assignments, some of which appear to be very similar, but with some differences in emphasis. The most common instructions are:

- *Compare*: this is asking you to identify and comment particularly on features in common, but also aspects of difference between two or more identified elements.
- *Contrast*: this is also a form of comparison but focusing predominantly on the areas of difference. This may be included with compare, as in: 'Compare and contrast ...', thereby emphasising that both the similarities and the differences must be brought out with equal weighting.
- *Describe*: is asking you to write a detailed description of the subject.
- *Discuss*: is often used to mean the same as describe, but it can also be used in its more literal sense of discussing the merits or shortcomings of a particular statement or idea, as in the example given (Figure 7.2).
- *Explain*: this is often used in the context of making clear how or why something happens, for example: 'explain the process whereby the blink reflex habituates during repeated stimulation'.
- *Evaluate*: the strict interpretation implies some form of quantitative assessment of information, but this may commonly be used in the more qualitative context of evaluating an idea.
- *Give an account of*: this usually has the same meaning as describe.
- *Write an essay about*: again, this typically means describe.

Be aware that some assignment titles will include more than one element, for example 'Discuss how ... and compare ...'. It is important to make sure that both elements are given enough focus. The pitfall to avoid is drafting a very long first section and then suddenly realising that you still have the second aspect to deal with, so that it is tacked on, very clearly as an afterthought.

Once you have identified what basic type of assignment you need to prepare, check for some other key pointers that give you some specific details about the approach to be adopted. These may include the following:

- *Write an illustrated account* ... this is emphasising that you need to include diagrams or other forms of illustration;
- *Using named examples* ... make sure that you use specific examples that illustrate your arguments;
- *Draft a report for a committee explaining the processes for investigating* ...
- *With reference to the recent experimental evidence, discuss the process* ...;
- *With reference to the research literature, compare the model for ... with ...*.

These groups of titles clearly show the emphasis that the question setter wants you to take in preparing your assignment. For example, it is not just a factual description that is required for the last two examples—you need to make sure that you are using the experimental evidence or recent research papers to support your arguments. As you progress through your studies, reference to research evidence should become an automatic part of your work and may not be referred to specifically in the title. In the case of drafting a report, it is important to consider the nature of the readership and their initial level of understanding.

Finally, make sure that you are very clear about the breadth of the subject material required: look for the key aspect that specifies a group or range. For example, in an assignment on the endocrine system, the title might refer specifically to steroid hormones; a title about ecology might refer to a specific type of habitat; or a title about animal behaviour might refer to a specific group or groups of animals. Also remember to think about the readership/audience for the assignment: what is their level of expertise within the subject.

In the coming chapters we will look in detail at how to tackle key types of assignment and will consider a key common element, which is researching the topic.

7.4　Some common problems with planning

For each assignment you will be given a deadline for submission. Figure 7.3 shows an illustration of time planning for an assignment expected to take three weeks to complete.

Ideally, these stages would be allocated appropriate portions of time to ensure that your preparation is as smooth and straightforward as possible. The division of time won't be precise, but you do need to allocate time for all the important elements not just, for example, the drafting itself. Importantly, the timeline also includes a gap between writing a first draft and writing a second draft. This will ensure that you can be more objective in your reviewing and make significant improvements to the work (see Chapter 11 *Revising drafts and finishing touches*).

If you can allocate your time in this way, it is a reliable method for increasing your chances of completing an assignment on time and to an appropriate standard. However, in practice, the process of producing an assignment is rarely as smooth and straightforward as this. Some common problems faced by students (and academics) when writing are as follows:

- leaving things to the last minute;
- doing too much research;
- doing little or no planning;
- taking too much time over the writing.

Figure 7.3 Suggested allocation of time for assignment task

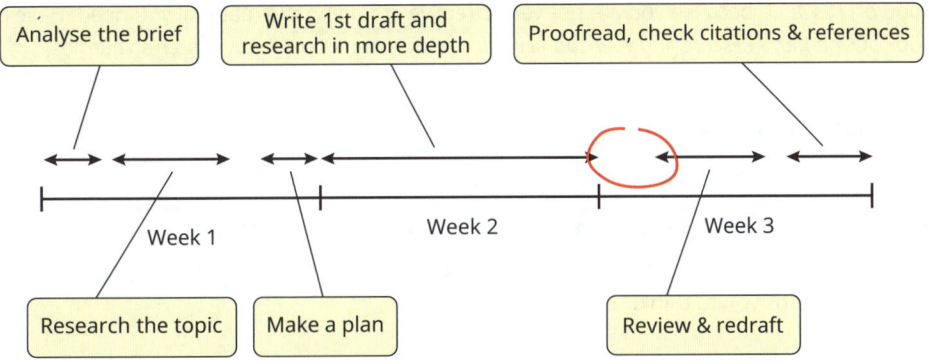

All of these are real problems, many of which you may have experienced for yourself. Each of the problems has a range of possible causes, and what causes the problem will determine what you need to do to address it. We will deal with each in turn.

7.4.1 Leaving things to the last minute

You might find yourself trying to complete your assignment at the last minute because:

- you have lots of other deadlines to meet;
- you feel anxious about coursework assignments and so avoid starting them;
- you just struggle to summon up the motivation to start earlier!

If you prepare an assignment like this, it means that you can be very focused and the work is often completed in the minimum time (because it has to be!). However, it can be a very stressful way to work and will probably affect the quality because you run out of time to review it and so end up handing in what is effectively only a first draft. If you find yourself in this situation, your immediate priority is to organise your time more effectively (as shown in Figure 7.3) so that you can allocate time to improving your work by reviewing and redrafting.

7.4.2 Too much research

You might find yourself undertaking too much research because:

- you get side-tracked by information that is interesting but not relevant;
- you do a lot of reading which generates a lot of notes;
- you make very detailed notes on what you read.

If you write an assignment like this it means that you have lots of material and you feel well informed; however, the volume of information can be overwhelming, not all of it is relevant, and there is less time available for other stages of the preparation process. Researching the topic is a vital task—nevertheless, you will not be judged on the information you have gathered, but on what you have submitted as the final piece of work. If you find yourself in this situation, your immediate priority is to develop a clear focus for your reading and note-making (Chapter 4 *What to do outside of taught sessions*). This will make you more efficient in gathering information, so you can then allocate sufficient time to the other stages. You may also be spending too long on research because you are not very effective at finding the material you need. Have a look at Chapter 9 *Researching your coursework topic* for some tips about search strategies.

7.4.3 Little or no planning

You might find yourself doing little or no planning because:

- you run out of time (see section 7.4.1 *Leaving things to the last minute*);
- you find planning a bit restrictive and you would rather just start writing, because writing helps you to think.

If you prepare an assignment like this, because the writing process helps you to think, it may mean that your writing will flow well and so it will read quite fluently. However, it can be a confusing and inefficient way to work because it is difficult to get an overview of the material and the logical sequence of the material can suffer as a result. If you find yourself in this situation your immediate priority is to set aside time for planning. The aim of planning is to provide you with an overview of the assignment structure so you can select and order your points in the most effective way.

7.4.4 Writing takes too long

You might find yourself taking too long over the writing because:

- you rewrite individual paragraphs again and again;
- you write too much, then struggle to cut it down to the word count specified;
- you simply find the process very difficult.

If you do work like this, it means that the drafting gets a lot of attention, which is a good thing. However, it can be an inefficient and frustrating way to work and won't leave you with enough time for other aspects of the process. If you find yourself in this situation your immediate priority is to speed up the production of the first draft, in order to have more time for reviewing and redrafting.

In this case, you need to have a plan of action. Start with a set of bullet points that summarise what you want to say, and then just add the content for each point. Don't try rewriting any sections until all the paragraphs are written. When you have the body of the assignment in place you will have a good appreciation of the overall structure. Now you can devote some time to editing or redrafting.

7.4.5 How to avoid common problems

You will have noticed from the common problems identified previously that crucial to improving your writing is managing your time. Many of us are not very good at time management. When we are set an assignment and told that the deadline for submission is not for six weeks, the temptation is to put it to one side and forget about it until a few days before the submission date. That means that many of us err towards being 'at the last minute-ers'. We all know that this can lead to problems.

So when should you begin working on an assignment? The obvious answer is 'as soon as you are given the title and your instructions'. It may not be practicable to race off and start preparing immediately, but you do need to incorporate your assignments into your course planning. Clearly, the sooner you start the better, but you may need to consider some other factors as well. For example, although you have to submit the assignment in week 8 of the term, you know that you won't finish covering the topic in teaching sessions until the end of week 4, so you are probably better off not trying to do anything until you have that background.

In Chapter 5, section 5.2, we summarised some useful principles to help you get organised and how thinking in terms of 'projects' can help. In section 5.4.2, we applied these principles to coursework, as follows:

- identify what you're trying to achieve;
- break it down into tasks;
- decide what needs doing by when.

One useful approach to your planning is to put all your assignments onto a single chart or diary (see Figure 5.1). For example, you could use a calendar. When planning, it is best to work backwards from the submission date and break the assignment down into its separate elements, as described earlier. This also has the advantage of giving you positive feedback because you can reward yourself as you complete each section on the calendar. If you make sure that you incorporate the submission dates for other pieces of work as well, then you can plan your time effectively, and also avoid hitting an unexpected log-jam of work that needs doing and handing in all at the same time.

When planning your timeline, you need to be realistic about allowing yourself enough time for the different components of the work; for example, carrying out research can be very time-consuming, particularly if you need to search for and read through (and understand!) a number of research papers. By the same token, you also need to be strict with yourself when undertaking the research, and not get carried away looking for more and more papers (see Chapter 9 *Researching your coursework topic*). Also, try to make sure that you factor in some time for proofreading before the submission date (see Chapter 11 *Revising drafts and finishing touches*).

Obviously, you won't be spending all your waking hours during those weeks working on the assignment: you will have other commitments, such as lectures, tutorials, and practical classes, as well as keeping time for socialising. Therefore, within your timeline, you will need to decide on what days you are going to focus on the assignment and for how long. Again, the key features are that you should try to be realistic, and then do your best to stick to your schedule.

What happens if things go wrong? Advance planning is again the key: whenever possible, leave a gap between your planned completion of the work and the submission date, then you have some leeway if something unexpected throws your plans out of line.

➕ Chapter summary

In this chapter we have had a broad look at the aims of different types of assignments with a focus on how to analyse the question you are set. We have also considered some of the common problems you may experience and reflected on the importance of planning. If you are experiencing these problems, then you will find it helpful to go back to Chapter 5 *Getting yourself organised*.

8 Creating written coursework

Introduction

In Chapter 7, we looked at the analysis and initial planning stages for preparing an assignment. We identified how important it is to understand what the aim of each coursework assignment is. In this chapter, we will explore the approaches to structuring your coursework assignment from the standpoint of the two aims we identified: discussion of a topic based on course materials and the research literature; and the presentation, analysis, and discussion of data you have collected. To do this, we will focus on two common assignment formats for each of these aims: essays (discussion of a topic) and practical reports (presentation, analysis, and discussion of data).

Although we will be looking at these two types of assignment in detail, the core principles we will be addressing are also relevant to most other coursework assignments that share these aims; for example, a presentation on data collected during a practical class uses the same content and order of information as a written practical report.

8.1 General guidance on appropriate language and voice for all assignments

The following points are relevant to the creation stage for all coursework assignments irrespective of their aim or format. It is important to use appropriate, scientific language and a style of communication that is professional, clear, precise, and objective. Adopting this is often an essential aspect of assignment marking criteria, where students can drop easy marks for poor quality writing. You should always bear in mind that the aim of scientific writing is to explain complex topics simply and clearly, so don't use long words for the sake of it.

8.1.1 Writing style

The style for scientific writing is normally impersonal and so avoids the use of personal pronouns such as 'I' and 'we' and the expression of subjective viewpoints. It also commonly involves using the passive voice rather than the active voice. The reason for this is that scientific audiences value 'objectivity'. That is to say, observations made during scientific research should be impartial, and not influenced by personal feelings or opinions.

For example, an introduction to an essay could be written as in Figure 8.1:

Figure 8.1 Personal and active

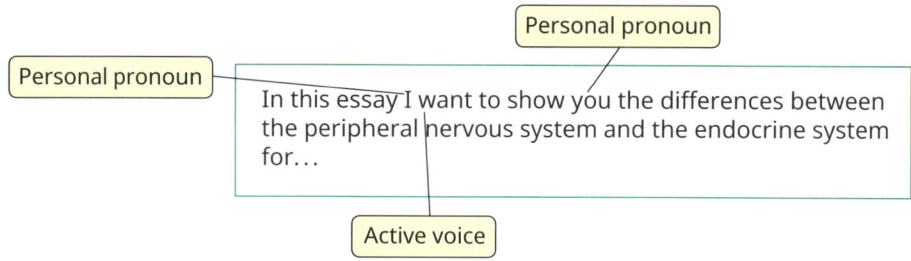

For a formal, scientific style (Figure 8.2), it would be better to write:

Figure 8.2 Impersonal and passive

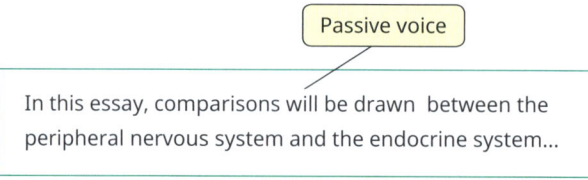

You should also avoid using contractions of words and expressing a subjective viewpoint. For example, when concluding on a topic in an essay or presentation you might write:

> The evidence I've given supports my feeling that …

This phrase contains the personal pronoun 'I' in the contraction 'I've'. It may be that, for particular reasons of emphasis, you wish to keep the word 'I' in this phrase, but you should avoid using any of the many contractions that appear in conversation such as 'I've', 'don't', 'couldn't', etc. Instead, you should write them out in full: 'I have', 'do not', 'could not'. [In this book we have chosen to use contractions because we feel that a more informal style is appropriate here, but it's (there we go again!) not appropriate for an academic essay.]

The phrase also expresses a very personal, subjective viewpoint: 'my feeling'. In the formal voice, this would be better written as:

> The evidence presented supports the argument that …

If it is important to express a personal summing up of the arguments presented, then you could write:

> On the basis of the evidence presented, I would conclude that …

8.1.2 **Scientific specificity and precision**

The use of generalisations often clouds the meaning of what is written or can indicate a lack of confidence on the part of the writer. For example, what does the following phrase actually mean?

> The peripheral nervous system quite often responds to stimuli ...

The phrase 'quite often' has no real meaning as its interpretation depends entirely on the reader. Relative terms such as 'quite' or 'fairly', or phrases such as 'In general terms ...', lack precision and should, wherever possible, be replaced with more precise phrases. Indeed, if you can, you should provide quantification of the statement, so rather than stating 'quite often' you can write:

> In 75% of cases the response was ...

As with the choice of words, precision in grammar is very important to ensure that our writing is clear and unambiguous. For example, the incorrect position of a comma in a sentence can change the whole meaning of that sentence as in the old joke about the panda that walks into a bar, eats something, pulls out a gun, and shoots the bar tender before leaving the bar. When questioned, the panda pulls out a book and points to the description of the panda which states it '... eats, shoots and leaves'. Had the description been phrased correctly as '... eats shoots and leaves', the bartender would not have been shot! There are numerous, entire books covering the subject of grammar (for example, *Eats, Shoots and Leaves*, by Lynne Truss; Harper Collins, 2009) so in this brief section we will look at just a few of the key rules. We consider grammar and syntax in detail in Chapter 11.

Having read this section, you should now be well aware that we have not written this book in a scientific style!

8.1.3 **Reference your sources**

Wherever you take other people's ideas from, you must reference them correctly in the text and list the sources in the reference list at the end of the report. Using an effective note-taking system when doing your independent research, as outlined in Chapter 9, will really help with this. This serves three important purposes:

1. Part of what you are being assessed on is your ability to independently investigate a topic, citations are part of the evidence used to assess this.

2. Insufficiently referencing your information sources is plagiarism (Chapter 9, section 9.2.1 *What is academic integrity?*). Referencing is dealt with in detail in Chapter 11, section 11.2.3 *Academic expectations*.

3. Focusing your assignment content on objective statements that are supported by evidence, thus reducing the use of subjective or imprecise points in your assignment, which would likely lose you marks.

8.1.4 Creating written assignments

This next section focuses on the general features of assignments that predominantly consist of long-form written communication; e.g. essays, practical reports, science communication pieces, policy-briefs, etc. That is not to say that they only use text to communicate key points: including figures, diagrams, and illustrations is a great way to efficiently convey depth of understanding of a subject within word count restrictions. Conversely, while visual and verbal formats do use text, there are different rules and restrictions for writing as part of a poster, infographic, or presentation; so, they are dealt with separately in Chapter 10.

Paragraph structure

The first stage for a written assignment is to create paragraphs based on the key points outlined in your plan. This section will provide guidance based on a plan made for a hypothetical essay assignment.

A paragraph is usually made up of several sentences that are linked together by a common theme. When writing, it often makes sense to have a paragraph for each key point outlined in your plan. You can edit and restructure your assignment once you have a first draft, so don't get too worried about structure at this stage, after all, it's often better to have a draft to edit than a blank page! The break between paragraphs can then be used to separate out different sets of arguments to help structure the logical flow of the assignment as a whole.

For example, if you want to make four main points in your assignment, then give each point its own paragraph. Each paragraph you write for the hypothetical essay might look like this:

- introduce the main idea (topic sentence);
- explain the idea (amplify the topic sentence);
- present supporting evidence or examples (quotation, study, expert opinion, or report);
- comment on the evidence (show how it relates to the main idea);
- conclude the main idea (link to the title or link to the next point).

A brief sample paragraph from our essay on the peripheral nervous and endocrine systems could be structured as in Figure 8.3:

Figure 8.3 Paragraph structure

Introductory sentence, setting out the theme of the paragraph	The speeds of communication of information are very different for the peripheral nervous system (PNS) and the endocrine system. In the case of the PNS, the conduction velocity of the action potential along the individual nerve axons can range from speeds of $0.5\ \text{ms}^{-1}$ up to $120\ \text{ms}^{-1}$. The differences in speed depend on the axonal diameter and presence or absence of myelin, with the highest speeds being generated by the mechanism of saltatory conduction of the impulses along large-diameter, myelinated axons. The transport of hormones takes place through the vascular circulation. Here, the speed of flow also varies, but in the arteries it ranges from 0.02 to $0.2\ \text{ms}^{-1}$. The differences in communication speed between the PNS and the endocrine system are important considerations when comparing the functional roles of the two systems.
Evidence to support the statement	
Final sentence linking back to the introductory sentence	

> **BOX 8.1 Paragraphs**
>
> There is no rule that prescribes how many sentences make up a paragraph, but the point where you are moving from one topic to another within the essay makes a logical place to insert a break and start a new paragraph.

Turning notes into writing

Regardless of when and how you plan your essay, you should support your points with evidence you have gathered together from your research. The inclusion of this evidence is a very important part of scientific writing. It is not sufficient to state:

> The endocrine system has widespread actions on the cells and tissues of the body.

Each statement should be supported by a specific example, so:

> The endocrine system has widespread actions on the cells and tissues of the body. For example, the thyroid hormones act to stimulate the metabolic rate of all the body's organ systems (Levy et al. 2005).

Note that the statement of the example is supported by a citation, indicating where the idea you have described came from (see Chapter 11, section 11.2.3 *Academic expectations* for more information on referencing). It is essential that, in your assignments, you communicate the ideas, concepts, and evidence found in your research in your own words. As we discuss in Chapter 9, section 9.7 *Can you spot plagiarism?*, if an essay or report were just a collection of other people's ideas it wouldn't score very good marks; in order to do well you need to engage actively with source material and express your own views by providing comment or analysis on your sources. There are many ways in which you can engage with source material. We are going to look at three examples:

1. Illustrating a point (as an example);

2. Providing evidence to support a point;

3. Providing evidence of a contrasting argument.

Precisely which method of engagement you choose is somewhat secondary; the important point is that you must engage with the sources you use in some way, rather than using material without comment or analysis. (Thanks to Prof. Raymond Dalgleish from the Department of Genetics, University of Leicester, for these examples.)

Illustrating a point

One way to use source material is to use it to illustrate a point. See Figure 8.4 for an example from the field of genetics. In this example, the idea is introduced, the evidence is summarised, and then an explanation is given of what this evidence illustrates.

Figure 8.4 Using sources to illustrate a point

| More than four decades ago, Marshall Urist began a series of experiments designed to characterize the factors necessary to induce bone formation (Urist, 1965). In 1975, he demonstrated that the 'bone morphogenetic factor' was part of the protein component of bone, rather than being associated with the hard mineralized fraction (Urist *et al.*, 1979). However, he was initially unable to isolate the factor as a single protein. This illustrated that the induction of bone growth might be controlled by several co-acting protein factors. | Introduction

Summary of evidence

Explanation of what this illustrates |

Note that this figure reinforces the guidance on structuring paragraphs noted earlier, where we can see a clear statement of a key point, the evidence that supports that point, and then a sentence that links the paragraph back to the overall topic being discussed.

Providing evidence

Another way to use source material is to use it to provide evidence. See Figure 8.5 for an example, again from the field of genetics. In this example, the idea is introduced, the evidence is summarised, and then the evidence is explained.

Again, note the way this paragraph is structured to align with the prior guidance. The final sentence that explicitly states how this information relates to the overall argument being developed is absolutely critical. Not including this critical 'so what' sentence is a common error in student writing.

Contrasting arguments

A third way to use source material is to use it to contrast arguments. See Figure 8.6 for an example, again from the field of genetics. In this example, the idea is introduced, some evidence is paraphrased, this evidence is criticised, contrasting evidence is then introduced, and finally the writer's opinion on the issue is offered.

Note this final paragraph provides an excellent example of critical thinking, an aspect of assignment grading that becomes more important as you progress through their studies.

Figure 8.5 Using sources to provide evidence

| Over several years, Urist continued to characterize his bone morphogenetic factor, making improvements to the purification procedures (Urist *et al.*, 1984). He showed that demineralized bone could be fractionated to yield low molecular weight protein morphogenetic components, but none of these was capable alone of inducing bone formation (Urist *et al.*, 1987). These studies suggest that bone morphogenesis is a complex process requiring the interaction of several factors. | Introduction

Summary of evidence

Explanation of evidence |

Figure 8.6 Using sources to contrast arguments

For many years, the precise identity of Urist's bone morphogenetic protein (BMP) remained unresolved except for the fact that it had a major component with a molecular weight of 18 kDa. In 2005, Behnam *et al.* demonstrated that the amino acid sequence of peptides derived from highly purified BMP show identity to regions of a previously characterized protein known as spp24. However, spp24 has a molecular weight of 24 kDa. It could be argued that Behnam *et al.* reached a mistaken conclusion because of failures in the strategy used to characterize BMP or that they simply failed to fully explain their findings in the published account of their work. Subsequent experiments, involving recombinant protein expression, have shown that the spp24 protein undergoes post-translational cleavage, yielding a product consistent with Urist's original 18 kDa BMP.

Introduction

Paraphrasing

Criticism

Speculation about the discrepancy

Resolution of the discrepancy

This paragraph specifically highlights uncertainty in scientific understanding of the topic, providing contrasting points on the topic from different perspectives and resolving them based on evidence from the literature.

These three ways of engaging with source material are not exhaustive; there are many others including agreeing, disagreeing, synthesising, reconciling, and developing. Some of these will be covered in more detail in the Chapter 12 on critical thinking. The important thing is that you provide some comment or analysis on the sources you use, rather than simply record them.

8.1.5 The writing process

Our general advice is to write in whatever way that allows you to have a first draft quickest; some people tend to spend ages trying to communicate a point in the 'perfect' way. This quest for perfectionism can result in a long time spent writing with little to show for it, which becomes disheartening and can mean that there is not enough time to engage with the other experiences that make up a university degree (not least the other assignments!). Just make sure you allow enough time and can also spend time on the next stage of the assignment creation process—proofreading and redrafting.

As you write each section of your assignment, make sure that you mark off each point you have covered and that the paragraphs link together, rather than appearing as completely independent sets of statements. There is no rule that states that you have to write the sections in the order in which they will appear in the final document. Some people find it is logical to write that way, but others prefer to write the body of the text, because then they have a clear idea of the key points that require communicating, and then the introduction and conclusion to provide the context. When writing research papers, many scientists prefer to write the methods sections of their scientific papers first (because that's the first part of the study they completed) and the abstracts last (because they need to summarise all of the bits they have already written).

Review and redraft

Reviewing and redrafting are important stages of the whole production process, but they often get overlooked because not enough time has been left to do them effectively. Again, this is where planning your use of time is so important: you should aim to leave yourself time after writing the original draft to review what you have written, rather than simply aiming to get the basic writing done by the deadline. Indeed, review is most effectively done when you have left a gap of a few days between writing the draft and undertaking the review. If you read through the essay immediately after you have written it, it is often difficult to look at it critically. See Chapter 11 for detailed guidance about the review and redrafting process.

Use feedback effectively

It is very tempting, when receiving a piece of marked work, to look at the mark or marks obtained, feel moderately satisfied (or disappointed), and then put the assignment away and not look at it again. It may well be that, since you submitted that assignment, you have moved onto another module and don't feel that any comments would be relevant to what you are doing now. If you do this, you cannot hope to improve the way you write because you are not paying attention to the guidance given as to where you went wrong or how you could have done better. Likewise, it is also important to know what things you are doing particularly well, so you can continue doing them.

Using feedback is a very important part of the learning process and we cover this in detail in Chapter 13.

8.1.6 Identifying your audience

A core aim of all scientific communication is to explain complex concepts simply and clearly (and not simple concepts complexly). Using simple terms should not mean that the assignment is simplistic, but it does mean that the content needs to be explained in such a way that readers can understand it without too much difficulty. As part of the assessment brief, there may be guidance on who the intended target for your assessment is. This is important because it influences the content and language you should include in your assignment. It is possible that the marking criteria will include how well you have aligned your content to the target audience. Some examples of target audiences and what that means are outlined here:

Specialist

If your readers are specialists, for example, your course tutors or other bioscientists, then you can assume a high level of knowledge. It is, therefore, acceptable to use technical language and terms in your writing.

Related

If the readership is from a field broadly related to the assessment topic, for example, if you are asked to prepare a 'policy brief', aimed at readers who are technically competent and educated but not familiar with the specifics of the topic covered, then you should be careful about using technical language and terms, especially without including clear definitions of what they mean.

General

If the target readership is from the general public, then you can only assume general knowledge. The people reading the article will probably have no familiarity with the discipline or subject area so you will need to avoid using technical language and terms, and should use only basic descriptions.

So, what might targeting language to the audience look like in an assignment?

Coursework aimed at a specialist scientific audience should make use of precise scientific terminology, which we can assume the audience is familiar with, to help communicate topics concisely.

> Figure 1. indicates that angiosperm alpha diversity was significantly higher in tropical forest plots than in temperate forest ones, but beta diversity was not significantly different.

The preceding sentence is appropriate for an assignment aimed at a scientific audience because we can assume that they understand the difference between alpha and beta diversity and the terms temperate and tropical. This means some very complex ideas can be communicated in a very short sentence (ideally supported with a graph of the data being discussed). However, the following might be more appropriate for a Science Communication style assignment aimed at educating an approximately GSCE-level science audience on a similar concept.

> A study led by Dr Greenthumb has found that forests closer to the equator, known as tropical forests due to their distribution between the tropics of Cancer and Capricorn, have more types of flowering plants than temperate forests (which are found in the highly seasonal regions covering most of Europe and North America). However, despite these differences in biodiversity (the number of species found in an area), both types of forest need similar sized protected areas to effectively conserve all the flowering plants found there.

This is a much longer piece of writing because we have had to take the time to explain many of the key terms to an audience that is unfamiliar with them. We also generally need to provide more of a focus on explaining what the results mean in terms that are familiar to the general public (such as conservation). Instead of relating this text to a complex graph that visualises patterns in data, you would be better off creating a diagram that effectively explains the difference between alpha and beta diversity.

All of this is designed to highlight that knowing your audience is an essential part of finding the right voice for your assignment. Complex terminology is helpful only when the audience is familiar with it, and should never be used just to make your assignment sound 'smart'. Similarly, only use a thesaurus to help you find a word that helps you more concisely explain a concept. Sometimes, you see sentences like the following in a student's writing ...

> The writer's grandiloquence and verbosity only served to obfuscate the veracity and adroitness of their reportage. The use of long, flowery words is not helping the reader to understand the point. As we said at the beginning of this section, the aim of scientific writing is to explain complex topics simply and clearly.

8.2 Writing essays

Essays are still a very common form of assessment in almost all degree courses in the biosciences, both as coursework and for examinations. They are the most common format of assignments designed to test your ability to independently research and coherently discuss a scientific topic. For many students, however, writing essays at university represents a real challenge as it is a skill that is only developed to a limited extent in pre-university courses. It may also be a long time (perhaps years) since you last wrote an essay, and that might have been for an English course, rather than a piece of scientific writing. This section is designed to help you appreciate what is expected of you, and guide you through the process of planning and writing essays. Before we begin though, consider this point: we often refer to the process of producing an essay as 'essay-writing', but how much of the time is actually spent writing?

Think back to the last coursework essays you produced: how much of the total production time was spent writing and what else was involved in the process? If you are like us, then your time allocation for essays is probably something like this:

- analysing the brief to determine what is expected of you (see Chapter 7);
- quite a few hours searching for information in textbooks and research papers, and even more time reading through the material trying to understand it!
- following that, you have to decide and note down what is specifically relevant and should be included;
- then some sort of planning stage, where you decide how to order the material, and how to begin and round off the essay;
- of course, there is the writing itself, which probably only represents about 20% of the total time spent working on the essay;
- finally, a relatively short time reading back to check the flow of the essay, that you had fully addressed the question and that there weren't any errors.

8.2.1 What are essays supposed to achieve?

This is probably a question many students have asked themselves, particularly during the early hours of the morning as a deadline is getting very close! It may also be a question that markers have asked themselves, too, as they wade through a pile of exam scripts or coursework essays. Actually, a well-designed and well-answered essay question involves the development and testing of a significant number of skills, some of which are not obvious. It is all too easy to focus on the scientific content of the piece of writing and not to think much about the way that content is presented and structured, but the skill of writing is a critical element in the production of a good essay.

We can identify a list of the skills being used to complete the task as follows:

- planning the use of your time;
- using feedback from previously marked assignments to produce a better piece of work (see Chapter 13);
- analysing the question (see Chapter 7);
- researching the subject matter (see Chapter 9);
- bringing together the information gathered from a range of sources, e.g. lecture notes, textbooks, research papers, web articles, etc.;
- planning the presentation of the information to address the question and make sure that all the key points are covered;
- ordering the material in a logical manner;
- framing the body of the text with an introduction, to set the scene, and a conclusion, to summarise the points made;
- writing succinctly and with clarity, paying attention to the grammar, syntax, and spelling;
- citing your sources appropriately (see Chapter 11, section 11.2.2);
- reviewing and proofreading what you have written (see Chapter 11).

This is a surprisingly long list and this book covers all of these skills, but the next section will focus on deciding the content of your essay and how to structure it. It is worth reiterating that this approach to structuring your assignment can be applied to most coursework that requires you to do independent research and discuss a topic or provide an answer to a specific question. Chapter 9 provides guidance on how to go about your research, and Chapter 10 provides guidance on how to turn the plan into visual (e.g. poster) or oral (e.g. presentation) formats.

8.2.2 Preparing your plan

In Chapter 7, we considered the importance of planning your time and analysing the question. So, by now you know what your lecturers expect from you. At this stage, there is often a great temptation simply to sit down, do some research, and start writing. This is, generally speaking, not a good idea: it is very easy to drift off the topic of the assignment, to leave important points out, or to produce a very disjointed piece of work. Planning is a very important part of assignment-writing and will help you gain the best possible marks by allowing you to produce a coherent structure to your assignment.

The exact timing of when you do the planning, however, will depend on your approach; there is more than one way of using plans. A plan for an assignment such as an essay is commonly used before writing of the first draft, but some people find it easier to start writing and then draw up their plan. Your preference will depend on how you approach the task of writing.

Consider the following descriptions and think which approach best describes your way of working.

Writer 1

'I like to make plenty of notes before starting. I like to have a good idea of what I am going to write before I begin the first draft. I find it difficult to begin a draft if I am not clear on the direction my writing is going to take.'

If this is you, then you need to use essay plans before you begin the draft, as per the conventional approach.

Writer 2

'I tend to make fewer notes than most people and tend to want to start writing as soon as possible. Often, I don't know what I want to say before I start to write, but I find that ideas occur once I have started. For me, most of my thinking is done through the process of writing.'

If this is you, then you need to set an early deadline for the completion of the first draft so you have plenty of time to review and reorganise the content. This is important because the content of your first draft will only be organised in the order in which the ideas occurred during writing. You will therefore need to spend more time on reviewing the draft than a person who has planned the structure beforehand. Your approach to writing can be very effective, but you must allow plenty of time to review and improve the structure of the first draft.

Writer 3

'I like to make plenty of notes beforehand and probably have a general of idea of the essay content before I begin to write. Once I have started to write I find that new ideas occur to me so the content of my draft is often different from my original intentions.'

If this is you, then you need to use essay plans before you begin the draft, as per the conventional approach.

Whatever your approach to the coursework creation process, it is likely that you will spend some time moving between planning, researching, and writing stages throughout the coursework approaches. However, it is always worth making sure you have some idea about what points you need to include to meet the assignment aims early in the writing process. So, make sure you have analysed the brief thoroughly before you start (Chapter 7). If you are unsure what the best approach to creating coursework is for you personally, an example strategy can be found in Chapter 9, section 9.4.1.

There are many different ways of drawing up plans. In this book, we will look at just two different formats: the linear plan and the visual plan. Let's assume you have been given the following title:

> Compare the endocrine and peripheral nervous systems as channels for control and communication within the body.

Linear plan

An example of a linear plan is shown in Figure 8.7. This is very useful if you like producing lists of key headings, to which you can then add sub-headings and specific details. Having drawn up the list of key points, you can easily reorder them to create a logical structure for the essay.

You can also see very quickly where you need to undertake research to provide the detail before you start writing. Furthermore, once you know how many key points you have, you can divide up your word allocation between the different points, with the target of giving a paragraph over to each point. When you do start writing, you can tick the points off as you go through them, and so make sure that you don't leave anything out. This approach also

Figure 8.7 An example of a linear plan for writing a coursework assignment

Introduction (approx 200 words)

Overviews of the endocrine and peripheral nervous systems
Descriptions of control - making things happen - and communication - sending information
Body of the essay (approx 1600 words)

- Mechanism of communication (200 words)
  - Endocrine - hormones transported via blood stream, slow, widespread action - e.g. thyroid hormone
  - PNS - action potentials conducted via axons, fast, specific to a locality
- Action at cellular level (300 words)
  - Endocrine - can be via specific receptors on cell surface triggering second messengers - e.g. adrenalin
  or crosses the cell membrane to act directly on the nucleus, e.g. testosterone
  - PNS - acts via specific receptors on the cell membrane - causes a change in the electrical potential of the membrane - e.g. acetylcholine
    - Point 3 (300 words)
      ○ ...
    - Point 4
      ○ ...

means you are not likely to drift away from the topic, which can happen if you just start writing without any planned sequence to the assignment.

Visual plan

An example of a visual plan for the same essay is illustrated in Figure 8.8. Here, the plan takes the form of linked ideas that are gradually developed into more specific details. This is often a very useful approach if the essay is conceptually difficult to put together and you are not sure how you can structure it. Once you have your basic plan drawn, you can allocate the numbers of words to the different levels so that each major branch is given appropriate weight. Again, as you write, you can tick off the branches that you travel along. For example, starting from the centre, you can tackle communication as your first topic. So you can write a paragraph comparing the routes by which the communication occurs, namely the bloodstream and the nerve axons; then moving to the outer branches, you can compare the characteristics of each means of communication. When you have completed this topic, you return to the centre point and start along the next main branch.

Figure 8.8 Developing a visual plan for an assignment

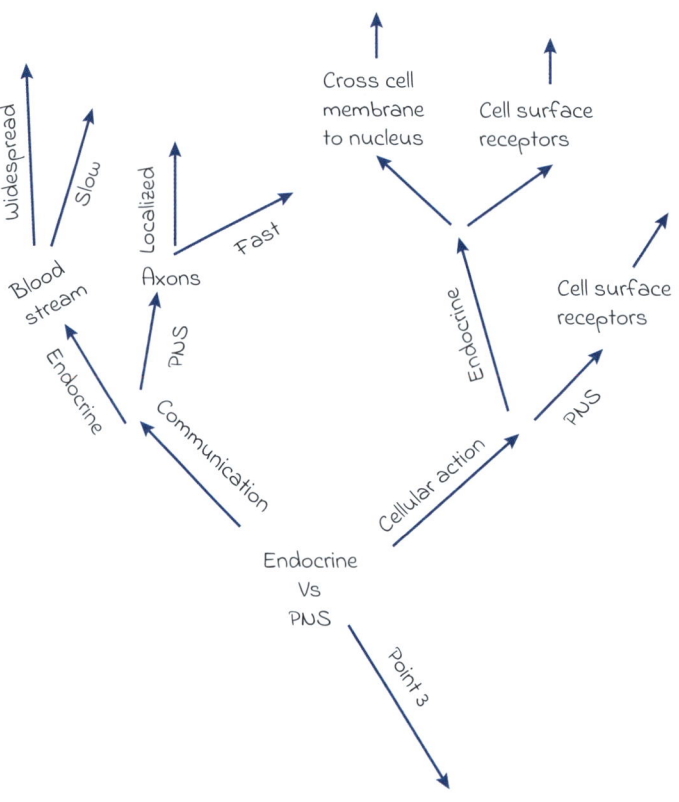

Try this—Produce an essay plan

Imagine that you are going to write an essay on your favourite hobby or interest (so that you don't have to do any research). Produce a plan for your essay using both linear and visual plans for comparison. Get someone to review it to see if they can identify the form of the essay and how it will flow. Think about which format came more naturally to you.

8.2.3 Structuring an essay

There are three main sections to an assignment such as an essay that you will need to write:

- the introduction;
- the body of the essay;
- the conclusion.

One way of summarising what these sections comprise is: tell your readers what you're going to tell them, tell them (in detail), then tell them what you've told them.

The introduction to an essay is used to frame the question and to set out what aspects are going to be addressed. An introduction is typically one or two paragraphs in length and should certainly not comprise more than 10% of the essay word count. The body of the essay is the bulk of the essay. In this section, you set out the arguments in detail and in a logical order. Finally, the conclusion is used to draw the reader back to the main focus of the question and to summarise the key points that have been made in the body of the text in a way that clearly addresses the question. As with the introduction, this should not normally exceed 10% of the essay.

Body of the essay

As you write each section of the body of the essay, make sure that you mark off each point on the plan and then make sure that the paragraphs link together, rather than appearing as completely independent sets of statements (see section 8.1.4).

Introduction

The introduction is a signpost for your reader, showing how you intend to answer the question. Remember what we said about the structure of the essay—in the introduction you are going to tell the reader what the essay is going to cover. One possible structure for an introduction could be:

- begin with a general point about the central issue;
- use the words of the title to show how you will focus on the question;
- indicate what the structure of your writing will be;
- make a link to the first point.

Note that in our consideration of the writing process, we have placed the introduction after the main body of the text. There is no rule that states that you have to write the sections in the order in which they will appear in the final document. Some people find it is logical to write that way, but others prefer to write the introduction after the rest of the essay, because then they have a clear picture of what it is they are introducing.

Conclusion

The conclusion is where you summarise for the reader your main points, linking back to the question and highlighting the most important aspects of your material. It acts as another signpost to your reader, rounding off the essay. As such it gives you the opportunity to:

- refer back to the question to show that your essay directly addressed the question set;
- summarise briefly for the reader the main points your essay covered;
- show the overall significance of the material in relation to the question;
- provide an overall assessment of theories or arguments, summarising your own viewpoint.

Again, you might want to write this in a different order from the final document: for example, you might like to have a conclusion written out first because you can use it as a guideline for reminding yourself where the essay is heading.

8.3 Writing practical reports

During your degree programme you will probably be asked to write up your practical work in a variety of different formats. These may range from simply filling in tables and writing short answers to questions listed on the practical schedule you have been given, to preparing a full report in the form of a scientific paper. In this section, we will take the scientific paper as the model as it covers the most common aspects of report writing. Make sure you are clear what format of report is expected from you before you write. For example, is it a full written scientific paper, or a scientific poster or presentation? Either way, the overall structure and content of the assignment will be similar but it will be communicated in different formats (see Chapter 10 for guidance on visual and oral communication).

It is important to remember that, while you will often need to work with a fellow student to conduct a practical experiment or undertake fieldwork, this doesn't usually mean that you are being asked to collaborate on the writing-up aspect of the report as well. Obviously, there will be similarities between two reports written by two students who had worked together to conduct an experiment, but these will largely be confined to the methods and results sections. However, in the discussion section, you will be expected to analyse and draw conclusions from your findings *independently* of each other. (See Chapter 9 on academic integrity.)

8.3.1 **Use an appropriate writing style**

The principles of writing for a practical report are the same as those for other pieces of scientific writing we have discussed earlier in this chapter. The key point to remember here is that you are writing for a professional readership and, therefore, you need to write in a formal, scientific style. We will consider writing up the practical work we did on the locust jump, as described in Chapter 3, section 3.3.

Avoid the use of 'I' or 'we'

Remember that scientific writing should be objective and impersonal:

> 'I measured the locust jumps at 10-minute intervals …' becomes: 'The locust jumps were measured at 10-minute intervals …'.
>
> Likewise: 'On the basis of these findings, we concluded that …' becomes: 'On the basis of these findings, it was concluded that …'.

Write in the past tense

You should describe what *was* done or what *was* observed. The only exceptions to this are when you are discussing illustrations or data presented in the report and the conclusions you can draw from them. For example: 'Figure 1 *shows* that jump distance *increased* with body mass …' or 'The key conclusion of this report *is* that jump distance *increases* with body mass …'.

Reference your sources

Wherever you take other people's ideas from, you must reference them correctly in the text and list the sources in the reference list at the end of the report. Remember that not to do so is plagiarism (see Chapter 9, section 9.2). Referencing is dealt with in detail in Chapter 11, section 11.2.3.

Use SI units

Measurements should normally be presented using the Système International d'Unités (SI) conventional form. The three fundamental units are the kilogram (kg), metre (m), and second (s). The SI system of units is based on seven 'base units', which are the fundamental measurements from which the other units are derived (Table 8.1). From these base units we derive a series of units for different measurements. Table 8.2 lists some of the most common units that you may come across in practical work and their definitions.

Note that when writing compound expressions, e.g. $m \cdot s^{-1}$, you should use the middle dot to separate the terms. This avoids confusion where terms could be misinterpreted; otherwise, for example, N·m could be mistaken for nm (nanometre) when handwritten.

Table 8.1 The seven base units of the Système Internationale d'Unités (SI units)

Quantity	Unit	Symbol
Length	metre	m
Mass	kilogram	kg
Time	second	s
Temperature	Kelvin	K
Amount	mole	mol
Current	Ampere	A
Luminosity	candela	cd

Table 8.2 Common units and their definitions

Measure	Unit	Symbol	Derivation
Area	square metre	m^2	
Volume	cubic metre	m^3	
Volume	litre	l	$0.001\ m^3$ $1\ dm^3$ (1 cubic decimetre)
Velocity	metre per second	$m \cdot s^1$	
Acceleration	metre per second per second	$m \cdot s^2$	
Force	Newton	N	$kg \cdot m \cdot s^2$
Energy, work	Joule	J	$N \cdot m$ $kg \cdot m^2 \cdot s^2$
Power	Watt	W	$J \cdot s^1$ $kg \cdot m^2 \cdot s^3$
Concentration	mole per cubic metre	$mol \cdot m^3$	
Temperature	degrees Celsius	C	K
Electric potential difference	Volt	V	$W \cdot A^1$
Electric resistance	Ohm	Ω	$V \cdot A^1$

Note that when writing compound expressions, e.g. $m \cdot s^1$, you should use the middle dot to separate the terms. This avoids confusion where terms could be misinterpreted; otherwise, for example, $N \cdot m$ could be mistaken for nm (nano-metre) when hand-written.

8.3.2 **Get the structure right**

The standard format of a scientific paper is as follows (see Chapter 4, section 4.2.2 for more detail):

- Title
- Abstract
- Introduction
- Methods
- Results
- Discussion
- References

You may also wish to include some additional sections, such as Acknowledgements, Appendices (see *Results*), and Keywords or a list of abbreviations used.

Of course, when writing up, make sure that you follow the instructions in your assignment brief as your lecturers may want you to focus on specific aspects of the report.

We will look at each of these sections in turn and you may find it useful to refer to some of the papers from your reading lists to look at the way these sections are organised and presented.

Title

The title must be brief and informative, so that the reader can see at once what the paper is about. In the case of scientific papers for publication, the title is very important as it will be used as part of the information for electronic searching, so authors will often include keywords to increase the probability of the paper being picked up by an electronic search.

Often, journals will limit the number of characters that can be used in the title. A typical limit would be 120 characters, including spaces, so it is good practice to write your titles within this limit.

Have a look at the four sample titles that follow, all of which describe the same investigation.

1. An investigation into the relationship between the body mass of an adult locust, of the species *Schistocerca gregaria*, and the maximum distance the animal can jump under non-restrained conditions.

2. The relationship between body mass and maximum jump distance for the adult locust, *Schistocerca gregaria*.

3. The locust jump.

4. Is bigger really better? Testing the jumping ability of different sizes of locust.

Which title provides the key information most efficiently?

The first title is unnecessarily long at 200 characters, although it does contain all the essential information. Here, the first section of the title 'An investigation into . . .' (or equally 'A study of . . .') is redundant and can be deleted without changing the sense of the title. The third title is too brief since it does not give any indication of the aspect of the jump being investigated. The second title is 108 characters long and contains all the essential detail for the reader to know whether to read further or not. Slightly humorous titles, such as the fourth example, may well be appealing for certain reports, and are often styles used by the popular press, but this approach should be treated with caution! A commonly used approach to writing titles of scientific papers is to concisely indicate either (a) the key research question the study is designed to answer, or (b) the key finding that the study data supports.

For example:

(a) What is the relationship between body weight and jump distance in adult locusts, *Schistocerca gregaria*?

(b) Jump distance is not correlated with body weight in adult locusts, *Schistocerca gregaria.*

Abstract

There is a real art to writing a good abstract. The abstract must provide the reader with all the essential information from the paper within a very limited number of words, typical abstracts being around 150–200 words in length. In many cases, the abstract may be used as a freestanding source of information: for example, when you are researching for an essay, you may well only read the abstract of a paper, rather than the whole paper.

The abstract will normally be the last part of the paper that you write and it is worth going through the following checklist to make sure you have covered everything.

- *Background*: very brief, but enough to set the context for the ...
- *Aims*: the key question you set out to answer using the ...
- *Methods*: again, these should be described very briefly, summarising the technique(s) used to obtain the ...
- *Results*: here, you need to present the key findings of the investigation, including the actual values obtained from which you drew the ...
- *Conclusions*: the key points you drew from the results. The conclusions should be related back to the initial aims.

The results and conclusions are the most important part of the abstract because, in a scientific paper, this is where you are presenting new findings. These two aspects should, therefore, make up the bulk of the abstract, with the background, aims, and methods being written as succinctly as possible. When you read through the abstract of your report (or better still ask a friend to read it through), the two points that should be immediately obvious are: what was found and why it is significant.

Compare the information content of the two sample extracts from abstracts describing our locust jump experiment and identify the key pieces of information that are missing from the second extract.

Abstract 1

The capacity of adult locusts to jump long distances to avoid predation is well known; however, it was unclear whether there is a relationship between the body mass of the adult locusts and their overall jumping performance. To investigate this, 20 male, adult *Schistocerca gregaria* (mean mass 1.6 ± 0.3 g) were stimulated to jump by means of a standard visual stimulus and the mean distance measured for ten jumps. A minimum of 15 minutes was allowed between each stimulus presentation to avoid habituation. The overall mean distance for the jumps was 0.95 ± 0.07 m and there was no correlation between body mass and mean jump distance ($r^2 = 0.15$). The findings indicate . . .

Abstract 2

Adult locusts can jump long distances. In these experiments the jumping ability of adult locusts was measured. The adults all jumped in response to a stimulus and the jump distances were averaged. Sufficient time was left between the jumps so that the response did not habituate. The results did not show any relationship between the mass of the locusts and the distance they jumped. The findings indicate . . .

Although the second abstract does describe the conclusion of the experiment, it does not give any indication of the initial aims. Nor does it provide any of the key information and numerical data to allow the reader to appreciate what was done and what was found.

As with the title, the abstract of a paper is commonly used as a means of electronic searching and so should contain keywords that identify the nature of the study and its conclusions. It is not normal practice to include references within the abstract.

Try this—Write an abstract

Go to the Wikipedia pages for either animal navigation, plant hormones, or cell communication (or another biological topic that interests you) and write an informative abstract that would help future visitors to the page.

Introduction

The introduction is the section of the report in which you describe the background and aims of the study. You might like to think of this section as addressing four questions:

- What has been done before?
- What still needs to be resolved?
- Why is it important to resolve this question?
- What do I aim to do?

A useful approach to drafting the Introduction is to start by setting out the aims of your study as a series of bullet points. Keep this list of aims in front of you while you are researching and writing the background to the study so that you keep clearly focused on the topic and don't start writing about sideline issues.

For a report in the form of a scientific paper you will need to read published papers on the topic of your investigation so that you can write a brief overview of the research that is underpinning your study. The skill in drafting this part of the introduction lies in writing a succinct summary of the findings of the previous authors and of the experimental evidence supporting those findings. As with all such writing, it is important that you present the material in your own words and fully reference the statements (see Chapter 11, section 11.2.2 on referencing).

The background text is normally presented as a factual account without discussion of the ideas presented in those papers. In most areas of research, there are areas of knowledge that are accepted as agreed and which you can describe as the background to the study. Developing from these, there will be areas where there is still disagreement, or questions remaining to be answered. In the case of disagreements in the published research, you will need to present both sides of the argument. This is best done simply by presenting your synopsis of the two arguments in sequence. Thus you might write:

> Derby and Joan (2020) concluded, on the basis of laboratory observations, that the locust jump is a behavioural response triggered by the presence of an attractive mate. However, in a more recent study undertaken in the natural environment, Bonnie and Clyde (2024) reported that the locusts only jumped to avoid being predated.

In the case of addressing questions still to be answered, you might phrase your statements as:

> Derby and Joan (2020) concluded, on the basis of laboratory observations, that the locust jump is a behavioural response triggered by the presence of an attractive mate. However, it is unclear as to whether this pattern of behaviour would occur in the natural environment.

Your description of the background to the study should, therefore, explain the scientific basis of the study, and provide an explanation of what is still not understood and why it is important that it should be understood. On that basis, you can then explain the aims of the study you have undertaken.

It is inevitable that, for most of your practical classes, you will not be aiming to resolve current controversies or discover the solution to a specific question in science, although it is perfectly possible that you may be doing that in your final year research project. For the most part, your practical work will be undertaking experiments that have well-known results. Despite this, it is important that you research the background to the study and present it as a synopsis, in your own words, of the scientific knowledge and can set out the aims of your experiment based on that knowledge. It is most likely that, if there is uncertainty in what the result will be in your study, it will be due to whether the expected result is found in the specific time, place, context, and study system you are working on. For example, the predictions may be well established for a species of locust but not necessarily the one you are working on. If this is the case, you can point it out in your introduction.

Methods

The methods section of the report is where you describe:

- what you did;
- how you did it;
- what you used to do it.

If your report is on a piece of fieldwork, you will also describe:

- where you did it;
- when you did it.

The methods section should be written in such a way that another scientist could repeat your work on the basis of your descriptions. Particularly when writing the methods section for your final year project, you may need to include more detail than would normally be found in scientific papers, for example, regarding the composition of solutions used.

Think of this section in terms of preparing a recipe for baking a cake. When you read a recipe, you need to know:

- what ingredients to use;
- how much of each;
- how to mix them together;
- what type of baking tin to use;
- what oven temperature to bake it at and for how long;
- how to analyse the results (eat it!)

Although you are describing what you did, you should still remember to write in the passive voice: 'The locust jump was measured using …' rather than 'I measured the locust jump using …'.

Check how much you need to write

In terms of reporting on class practical exercises, the necessity for a methods section is variable. For example, where you have been given a detailed schedule to follow, you will quite probably be told that you do not need to copy out the methods again. Even in this case, however, you will be expected to report on any changes made to the protocols, particularly since these may affect the results you obtain.

Assuming that you will be writing a full report, you will need to describe the methods you used in detail. If you are using a new technique, or have been instructed to give full details of the protocols, then you should set them out in full. In most papers, however, if you are using a standard procedure that has already been described in detail in another paper, then you can simply refer to the previous paper and need only specify any changes you made:

> The locust leg muscles were stained for ATPase using the method of Peters et al. (2016), but with an extended incubation period of 10 minutes.

Species names

If you are using animals or plants in your experiments, you should give the full name of the species, which is written in italics for the genus and species name, and the number of individuals involved. Where it is significant to the experiment, you should also give other details, such as the age, sex, and physical dimensions (e.g. height, weight). Thus, for the locust jumping experiment, your methods section might state:

> The jumping experiments were performed using 20 adult, male locusts (*Schistocerca gregaria*), mean weight 1.6 ± 0.3 g (standard deviation).

Human subjects

If your research project involves human subjects, you will need to have ethical approval from a local or NHS ethics committee for the work and you should quote the approval in the methods. You will also need the informed consent of the subjects and this should also be stated. Your project supervisor should advise you on the procedures in your university.

Chemicals

When listing chemicals or other substances that were used you should give the name of the chemical in full (e.g. sodium bicarbonate) and/or its chemical abbreviation (e.g. $NaHCO_3$), along with the concentration of a solution and the amount used. If you have used specialist reagents, you may also need to note additional details, such as the name of the supplier.

Specialist equipment

When describing the equipment used, again the amount of detail you need to give depends on the degree of specialisation. For example, if you are stating that you used 3 g of sodium bicarbonate, you don't need to give details of the balance used to weigh out the chemical. However, you might have used a specialist piece of equipment for your experiment, for example, a confocal microscope or a specific computer interface for data logging. In this case, you should give the details of the equipment, such as the manufacturer and the model number. In some cases, it may also be helpful to include a simple diagram of the apparatus used, to show how different pieces of equipment were connected together.

Statistics

We are not going into any detail of statistical analyses as there are many books for the Biosciences that address this topic. However, there are some key points that you must remember.

For many experiments you will be undertaking several measurements. You will, therefore, need to give descriptive statistics in your methods. The basic statistics would be measures such as the *mean*, the number of measurements taken (*n*), and the range of the values. For example, if you quote a mean value for a result, you need also to give the number of tests carried out (e.g. $n = 27$) and an indication of the spread of the data. The spread of the data is often given in the form of the *standard deviation*, which gives the reader a measure of the confidence that can be placed in the mean value. In the methods, you would state:

> Means are given with the standard deviation ... the adult locusts used had a mean leg length of 21 ± 3.2 mm (n = 20).

You may also be employing statistical tests in your analysis of the data, for example, to distinguish whether a treatment had an effect on the sample population. For most practicals, you are likely to employ routine tests, such as the Student's t-test, the Mann–Whitney U-test, or the chi-squared (χ^2) test. In such cases it is sufficient to state the name of the test and the level of significance that you are accepting, e.g. $P < 0.05$. For example, in the case of comparison of two groups of locusts you might state that:

> The mean leg length of the adult locusts was significantly greater than that of the juveniles ($t = 3.2$, $P < 0.05$, Student's t-test).

More extensive details regarding statistical analysis can be found in *Research Methods for the Biosciences* (Holmes et al., 2016).

After all that

When you have completed writing the methods section, read through it and check that you could repeat the experiments from the description you have written (think back to the recipe—could you bake the cake?)

Results

The results section is the core of the report. Here, you *describe* and *show* what was found along with the analyses of those findings. You should aim to lead the reader through the findings, highlighting the important features. You do not, however, attempt to interpret or explain the results in any way—this section is a factual description.

When writing the results section, you should use your data to tell a story. It is very important that the text you write can stand on its own in terms of communicating the key points. The graphs, tables, and other images are then used to support the description. A common error in students' results sections is for the data simply to be presented as the graphs and tables with little or no explanatory text, leaving the reader with the task of trying to work out what is important. So, rather than just writing:

> The results for the jumping experiment are shown in Figure 1.

It is much better to write:

> From Figure 1 it can be seen that there was no significant correlation between the distance jumped by the adult locusts and their body mass ($r^2 = 0.15$).

In the case of the first extract, it is left to the reader to draw their own conclusions from the plot shown in the figure. By comparison, the second extract provides the key summary of the observation, using the figure to illustrate the point and providing statistical confirmation, by means of the coefficient of correlation, that there was no relationship between the two variables.

Calculations and levels of accuracy

When using calculators and spreadsheets, be aware of the levels of accuracy of your measurements; often, a calculation performed on a calculator can result in a number with a string of up to 10 digits. For example, when measuring the jump distance for our locusts we might only be able to measure the distance to the nearest centimetre. So an individual measurement might be recorded as 0.94 m. However, the mean calculated for all the jumps, when using a spreadsheet without limits on precision, might give a figure of 0.942857142 m. If you put this number into your results, it implies that you could record the jumps down to a precision in the nanometre range! As a very simple rule of thumb, don't present calculated values to any more decimal places than the original measurements.

What is the best way of presenting the data?

This is often a question students find hard to answer. There is a strong temptation to adopt the 'everything' approach—if you include absolutely everything then you can't have forgotten anything! However, for the reader, this can make interpretation difficult and someone marking your report may conclude that the reason for putting everything in was because you weren't sure what was important and what was not.

Remember that you are telling a story, or answering a specific research question, and so try to identify what the key points are, in sequence. Where possible, you should present your data in summary form, rather than long lists of raw measurements (which can be included as an Appendix at the end if necessary). If you have measured 20 jumps for an individual locust, the reader doesn't want to see a table of the individual jump lengths. Instead, you can simply present the data in summary form as the mean, the standard deviation, and the number of measurements.

This statement provides the reader with as much effective information as does the full list of individual measurements and is in a much more digestible form. For a practical report, you may wish to include all the raw data, in which case you can add it as an Appendix to the end of the report.

> The mean jump distance for the adult locust was 0.87 ± 0.02 m ($n = 20$)

The three main ways of presenting data are as follows:

- text;
- tables;
- graphs.

Text

Use statements of specific values in the text when you are referring to only one or two items, so, the example used earlier, stating the mean jump distance, is appropriate for presenting in text.

Table 8.3 Measurements of the body mass, femur length, and the maximum jump distance for five adult locusts

| Give your table a number so you can refer to it from the text | Table 1 Measurements of body mass, femur length, and maximum jump distance for five adult locusts | The title should allow the reader to understand the table on its own |

Table 1 Measurements of body mass, femur length, and maximum jump distance for five adult locusts

Locust	Body mass (g)	Femur length (mm)	Maximum jump distance (m)
1	1.34	21.4	0.96
2	1.53	21.8	0.87
3	1.65	21.9	0.88
4	1.72	22.0	0.92
5	1.68	22.0	0.97
Mean	1.58	21.8	0.92
SD	0.15	0.25	0.04

Each column must have a heading to identify the data set and the units of measurement

Tables

Tables are very useful for presenting data in an organised form, particularly where you wish to present several variables together to give an overview of a set of results. The layout of the table (e.g. Table 8.3) is important to ensure that the contents can be easily read and understood. In particular:

- *try to avoid having too many columns*: having more than five or six columns can make comparison of the data items difficult and also means that the text may be compressed to fit on the page;

- *make sure the rows and columns have clear headings, with units*: so that the nature of the data is immediately obvious;

- *spread the data out* and have clear delineation between rows and columns.

Graphs

There are many different ways of presenting data as graphs. The types of graph you are most likely to use are:

- bar chart;
- histogram;
- scatter graph;
- line graph.

Bar charts can be used to display data related to frequency. So, if instead of displaying the relative proportions of each grouping of locust, you wanted to display the actual numbers in each group, you would use a bar chart as shown in Figure 8.9.

Note that, in Figure 8.9, each axis is labelled to identify the information being displayed and that the scale of the *y*-axis (vertical) is set appropriately to display the spread of the data (have a look at the Appendix at the end of this chapter for some examples of how *not* to draw graphs). A key feature of the bar chart, unlike the histogram (see next paragraph), is that the

Figure 8.9 A bar chart showing the numbers of locusts divided approximately in terms of age, based on exoskeleton colour (pink: youngest; brown and yellow: oldest), in the sample tested by the class (n = 100)

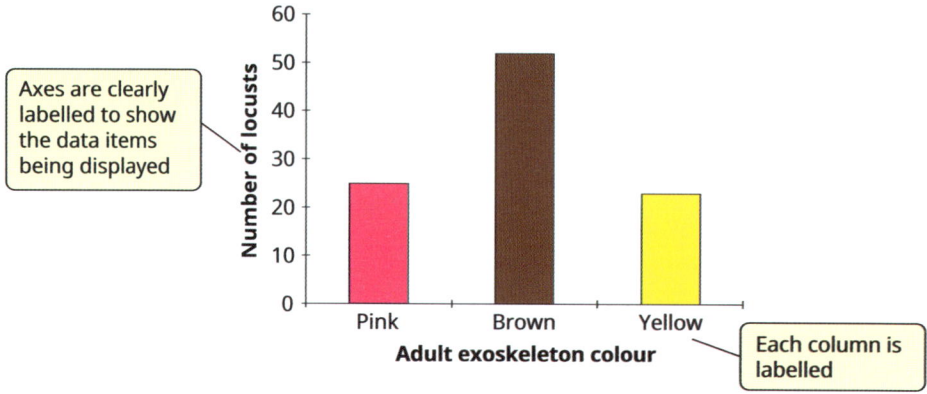

Axes are clearly labelled to show the data items being displayed

Each column is labelled

bars are separated by a space. This reflects the fact that the *x*-axis (horizontal) is displaying discrete items and not parts of a continuous population. The horizontal axis is just there as a platform for the bars.

Histograms are also commonly used to display data related to frequency, but in this case the variable normally shows a continuous variation. This means that the groups along the *x*-axis form part of a spread of data (e.g. the height or weight of a population), rather than being separate classes, as is the case for the exoskeleton colour. The example shown in Figure 8.10 illustrates the frequency distribution of body mass for the sample of 100 locusts tested in our class experiment. Here, the body masses of the locusts have been grouped into seven groups (or 'bins'). It is important to select the size of the bins carefully so that you can

Figure 8.10 A histogram showing the frequency distribution of body masses in the sample of adult locusts tested by the class (n = 100). Note that body mass shows a continuous variation and the divisions between the bins are chosen to best display the data

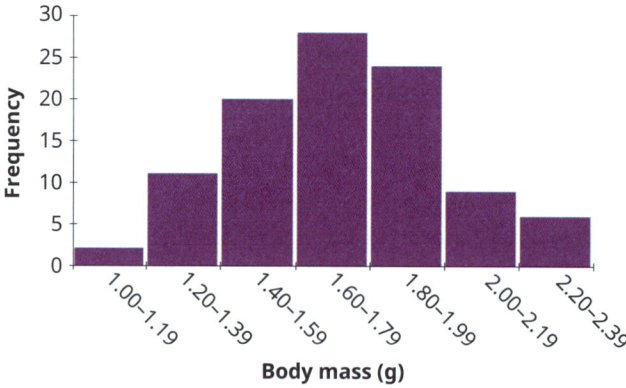

display the data effectively. Avoid having a large number of bins with only one or two items in each. Likewise, having too few bins may mean that the pattern of distribution of the data cannot be seen. You may need to try a few times before you get the best layout.

Scatter graphs are often used to display the spread of data for two variables that are related to each other. For our sample of locusts, this might be the relationship between body mass and maximum jump distance, as shown in Figure 8.11.

How do you determine which variable goes on the *x*-axis (the *abscissa*) and which on the *y*-axis (the *ordinate*)? The variable placed on the *x*-axis should be the *independent* variable. This means it is the variable that is already known, or can be controlled by the experimenter. In the locust jump experiment, we have a group of locusts of known body mass and we want to see how far they can jump. The body mass is the independent variable and the jump distance is the *dependent* variable, which is placed on the *y*-axis. The dependent variable is, therefore, the variable that may change as a function of the independent variable. So, in this experiment, we are proposing that the distance a locust can jump depends on its body mass. The logic of this should be clear: in this case, we can assume that the body mass of a locust does not depend on the distance it can jump!

When displaying data as shown in Figure 8.11 there is a further decision to make, which is whether to draw a line to indicate a trend in the data. For example, it appears that, although the data are spread out, there is a trend of longer jump distances being obtained by the larger individuals. Discussion of the processes of line-fitting is beyond the scope of this book: you should refer to the companion text on statistics (*Research Methods for the Biosciences* by Holmes et al., 2016). As a rule of thumb, however, be wary of the complex line-fitting programs that can be found in many spreadsheet and statistical software packages.

These programs can fit scattered data with very complex mathematical equations, which can be very difficult to interpret. If you think there may be a simple trend such as a linear relationship, then use the program to calculate the linear regression line that best fits the data. However, you should only draw conclusions regarding the linearity of the relationship if you also calculate the coefficient of correlation and this demonstrates a statistical significance.

Figure 8.11 A scatter graph showing the relationship between body mass and maximum jump distance for adult and juvenile locusts (n = 56)

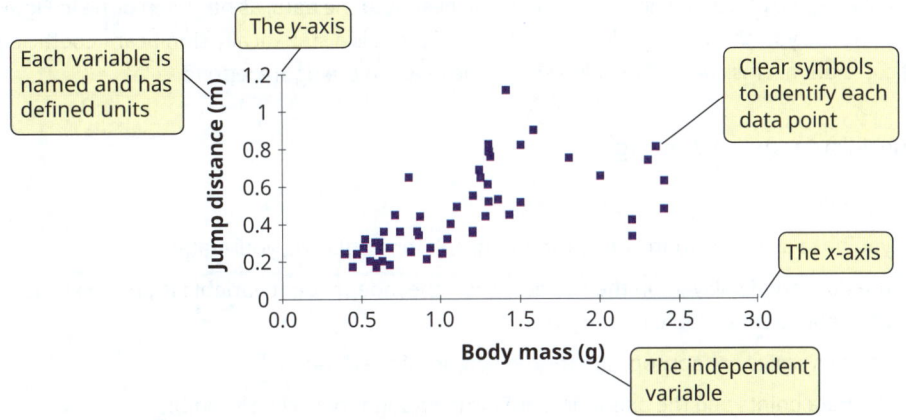

Figure 8.12 A line graph illustrating the relationship between the mean power output of the locust extensor muscle and the initial joint angle for 10 adult locusts. The error bars show the standard deviations of the means. There was a significant negative correlation between power output and joint angle ($r^2 = 0.91$, $P < 0.01$; method of least squares), as indicated by the linear regression line.

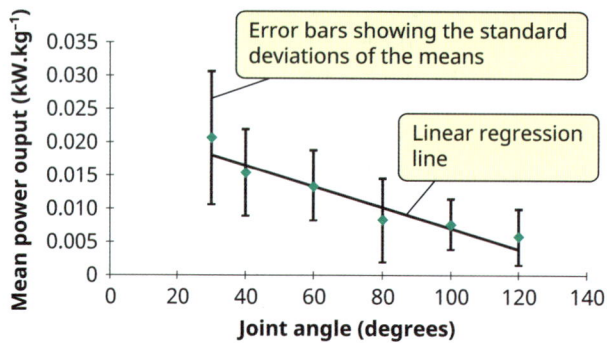

Line graphs: rather than having a scatter of data points, your experiment may have involved taking a series of measurements against a set baseline, as for the population in Figure 8.12. An example of this would be taking measurements at set time intervals. For such data, time is represented on the *x*-axis and the change in the variable on the *y*-axis. In the case of the data shown in Figure 8.12, the experimenters have measured the power generated by the leg extensor muscles as a function of joint angle. The measures have been repeated for 10 individuals and then averaged, so each data point actually represents the mean of the 10 individual results. This plot, therefore, illustrates another important feature of graph-drawing which is that, for averaged data, you should also indicate the spread of the data items contributing to each value displayed. This is done by including the standard deviation as vertical error bars above and below the data point. The reader can then see how reliable the plotted points are in illustrating the trends being shown.

As for the scatter plot, you will need to decide whether to draw a line to illustrate a trend in the data. This can be done either by simply joining the individual points together, or by calculating the regression line that shows the best fit to the data, as has been done in Figure 8.12, Again, you should only plot a best-fit line if there is a statistically significant coefficient of correlation. This coefficient value should be given in the figure legend.

Checklist for graph drawing

Make sure that:

- you have used the appropriate form of graph for displaying your data;
- the data are displayed on the correct axes—the independent variable is on the *x*-axis, the dependent variable on the *y*-axis;
- the axis scales are appropriate for the spread of the data;
- the data points and the axis labels are large enough to be clearly legible;

- the axes are labelled and the units of measurement given;
- the graph has a clear figure legend or title so the reader can appreciate what is being displayed without needing to read the accompanying text.

Look at the Appendix at the end of this chapter for some examples of how *not* to draw graphs.

Using images

You might use a variety of images, such as photographs, in your reports. These photographs may be taken using microscopes or electron microscopes, or they may be photographs of a landscape. The key rules here are to make sure that:

- the illustration has a legend or title that explains what is being shown and how the image was obtained;
- any specific aspects of interest are identified by arrows, so the reader can appreciate what is being described;
- there is an indication of the scale of the image—this is usually done by adding a labelled scale bar to the illustration.

Results checklist

After you have completed the writing of your results section, read it through carefully and ask yourself the following questions:

- Have I clearly described the key findings in the text?
- Have I given the correct units for all the measurements?
- Do the tables and graphs present the data in a clear form?
- Are the key conclusions supported by appropriate statistical tests?
- Have I avoided trying to interpret the findings?

Discussion

The discussion is the section of the report where you interpret the findings from your experiment and place them in the context of the research literature. When it comes to writing up a practical class or project, this is often the most demanding part of the exercise, because you must display your understanding of the work you have done and how it fits into the studies undertaken by other scientists. As we mentioned earlier in this chapter, if you have undertaken the practical work in a group, this is where you will be expected to analyse and draw conclusions from your findings *independently* of each other.

A useful starting point for the discussion is to write an introductory paragraph that summarises the key findings from the results:

In this investigation into the mechanisms of the locust jump, it has been shown that . . .

This paragraph then provides a good link with the results section and gives you a clear list of the points you need to interpret. There should also be a link back to the aims that you set out in the introduction. Having identified your key findings, you then need to discuss each of them in turn. As for the introduction, you will need to refer to the relevant research papers, highlighting their findings and relating them to what you have found. It may be that your findings are different from, or even contradict, those presented in other papers. In such cases you will need to explain why you think this may be so—don't always fall back on the staple explanation that you probably got something wrong! Whereas that may be true in many cases, there may also be genuine reasons for your experiments having generated different results. If you are writing up a research project, it is likely that you will have new data and these will need careful interpretation with reference to the research literature.

You should remember to round off the discussion with a concluding paragraph that summarises both the key findings and their interpretation. In class reports, you may also be encouraged to identify what experiments you might do next in order to progress the work. An example of a short conclusion for our locust jump experiment might read as follows:

> In these experiments, it has been demonstrated that there is no significant correlation between the body mass of adult locusts and the maximum distances that they can jump. These findings have been discussed in relation to the studies by Smith and Jones (2014), who reported that the processes of power generation by the adult locust could be more closely related to age rather than body mass. The relationship between …
>
> There was a large range of variation in the results obtained in these experiments and any future study should employ more locusts, with more jump trials for each locust, in order to increase the probability of demonstrating significant differences between the different groups. Furthermore, the failure of the heating system during the course of the experiment meant that the later jumps were being performed at temperatures that were sub-optimal for the locusts.

Reference

As with all your written work, you must include a full list of the references that you cited in the text (see Chapter 11, section 11.2 *Proofreading the document*).

Other sections

You may wish to include some additional sections to the report. The most common of these would be an Appendix and a set of Acknowledgements.

Although the results section is where you present your data, this is often in the form of processed information, for example, the means of sets of measurements. However, you may wish to include all the raw data, to demonstrate the full set which you have obtained and from which you have derived the summary results. You may also wish to include worked examples of your calculations, such as the statistical tests. Under such circumstances, these can be presented in tables in an Appendix. This is useful in a practical report or in your project report, but would not be typical practice in a research paper.

It is unlikely that you will need to include an Acknowledgements section in a report from a class practical, but this is more common in a project report. For your project you may have been given specific support by academic or technical staff, or been given access to specific items of equipment that would not normally be available to you. It is then a matter of good practice to acknowledge this help in a brief statement at the end of the report. This should not be treated like the eulogies at the Oscar ceremonies—just give brief, factual acknowledgement of any specific support you were given.

8.3.3 Write the sections in an appropriate order

Writing a report can often seem a daunting process, particularly if it is for a large piece of project work. Before you put pen to paper (or fingers to keyboard), check again any brief you have been given for the structure of the report, so you know exactly what is expected of you. Also refer back to any feedback you have received on previous reports.

Some sections of the report should be much easier to write than others. For example, the Methods section is simply a factual description of what you did and so this is often a good section with which to start.

The next section to tackle is the Results. You should think carefully about how you want to present the data first of all, and then produce your graphs and tables, and carry out any statistical analyses. When you have the data presentation sorted out, the process of writing is fairly straightforward since you are aiming to describe the results that you have in front of you. Don't forget that you are telling a story in which the storyline is illustrated by the figures and tables.

Having set out the key findings of the experiments in the results section, you are then in a position to consider their interpretation. Before beginning to write the discussion, however, you should read through the relevant research papers so that you have a clear picture of the relationship of what you have done to the current research. Take each of your key findings in turn and write about it, making sure you explain the relationship with the previous research in order to interpret what you have found.

The final sections to write are the Introduction and the Abstract. Having written the Discussion, you should be in a position to present the background to the work and to set out the aims. As for the Discussion, you will need to refer to the relevant research articles to support your statements. This should lead logically into the aims of the experiment and you can check off the aims against your final conclusions. The Abstract is the final piece to be written, based on a very brief synopsis of the other sections of the report and highlighting the key findings of the research.

Finally, as with all written work, it is always a good idea to put the completed report to one side for a few days and then to re-read it, checking that it all ties together and could be understood by someone who had not done the experiments.

Chapter summary

In this chapter we have considered the general principles of using appropriately scientific language and professional writing style when creating coursework assignments, as well as how to tailor content to specific audiences. We also looked at how planning your approach to writing is very important to ensure that you answer the question and cover all the points you need to.

The chapter has particularly focused on the structure of two of the most common assignment formats as examples of the most common assignment aims: the essay (research and discuss a topic) and the practical report (present, analyse, and discuss data).

The aim of an essay is to show that you can independently research a topic or questions, and turn that research into a coherent and well-evidenced discussion of the topic or answer to the question. This chapter has discussed some techniques required to do this well: develop an overall structure that orders the content logically and then create a series of paragraphs and sentences that communicates the key content points clearly and concisely. These techniques can be applied to other coursework assignments that require you to research and discuss a topic.

The aim of a practical report is to present, analyse, and discuss data (most likely data you have collected during some form of practical). Practical reports most commonly follow the structure of a scientific paper: Title, Abstract, Introduction, Methods, Results, and Discussion. This chapter has considered each of these sections and provided guidance on how to meet the expectations of each of these sections in your coursework. The guidance can be applied to any coursework assignment in which you need to analyse, present, and/or discuss data or a practical technique or methodology; however, you may need to provide more or less specific detail depending on the format of the assignment (e.g. written report versus poster).

For almost every assessment at university, you will need to undertake independent research of the scientific literature and make sure that you cite and reference your sources of information correctly. All while meeting the expectations of Academic Integrity. Guidance on this is provided in Chapter 9 *Researching your coursework topic*.

This chapter has focused on long-form written assignments that expect you to write in full sentences and paragraphs. While the guidance on structuring and content applies to all coursework types, specific guidance on how to create coursework that is more focused on visual (e.g. posters) and oral (e.g. presentations) communication is covered in Chapter 10 *Developing visual and oral presentation skills*.

Finally, we have advised you that you must always leave time for the review and revision process. Guidance on how to undertake this effectively irrespective of coursework type is provided in Chapter 11 *Revising drafts and finishing touches*.

 Appendix: examples of how not to draw graphs

The display of data in graphical form is a key part of the preparation of many reports of practical work, whether it be for enzyme assays or the populations of birds in a cliff nesting site. In the section on graphs, we have described some of the key principles in drawing graphs, but there are many pitfalls, especially when using spreadsheets or graph-drawing packages. Some of these pitfalls are more obvious than others. Have a look at the graphs that follow (Figure 8.13 and Figure 8.14).

Figure 8.13 shows a number of common errors (although rather exaggerated) that might be made when displaying the data shown in Figure 8.12:

- *The figure legend is too brief and does not explain what is being illustrated*. The test here is: if the reader only had the figure and the legend to go on, and no accompanying text, would they be able to understand what was being shown?

Figure 8.13 Power output against joint angle, locust experiments

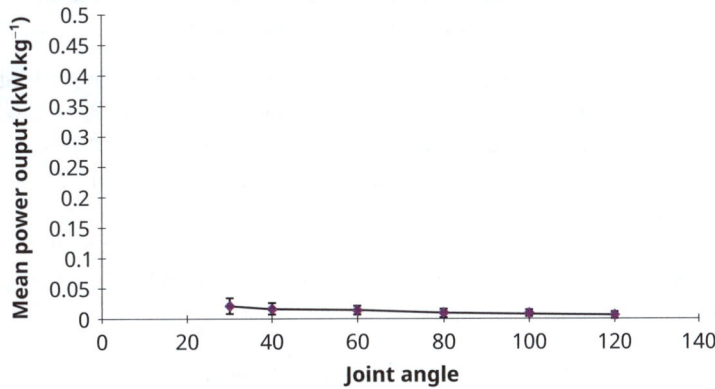

Figure 8.14 Scatter plot showing the relationship between body mass and maximum jump distance for adult and juvenile locusts, *schistocerca gregaria* (n = 56)

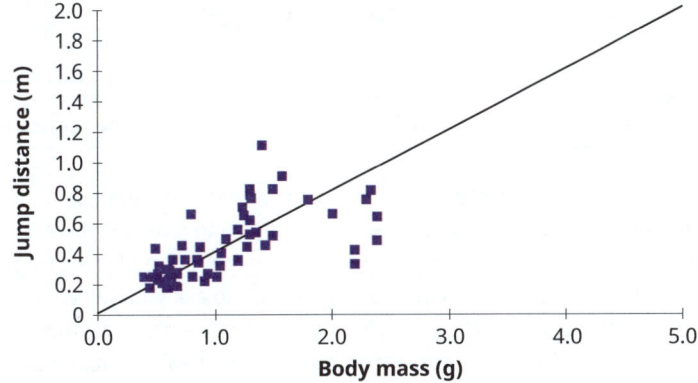

- *The scale for the y-axis is inappropriate.* The vertical spread of the data points does not even extend to the first scale point on the axis. Furthermore, the compression means that the error bars are not actually visible. The scales should be set so that the data points extend over almost all of the length of the axes.

- *The font size for the scale on the x-axis is too small to be legible.* Make sure that the fonts for the two axes are legible and are the same size. The *x*-axis label 'Joint angle' does not specify the units.

- *The numbers on the y-axis are not set to the same level of precision.* Make sure that the same number of decimal places is specified for the scale indicators.

Figure 8.14: At first sight this scatter plot might appear to be OK since the axes are appropriate and clearly labelled and the legend informs the reader as to what data is being presented. However, there are some key flaws in the presentation:

- A straight line has been drawn through the data points but there is no indication as to the basis for drawing such a line: was it simply drawn by eye as appearing to be the line that best fitted the data, or was it calculated using linear regression? The method for arriving at drawing such a line needs to be stated in the figure legend. If it is drawn by linear regression, then the method chosen for calculating the line needs to be stated and the coefficient of correlation given so that the reader knows whether the linearity is statistically valid. An example of the statement to include in the figure legend would be as follows:

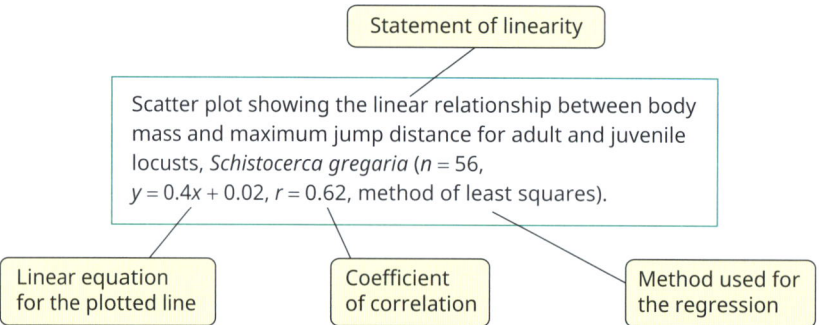

- Is it valid to extrapolate or interpolate a line? The straight line here has been drawn back to the origin (coordinate 0,0); this is *interpolation* and it makes the assumption that the linearity of the relationship continues below the range of the data set on which the calculation was based. Therefore, you need to ask yourself whether this assumption is justified or in this case biologically possible—after all, how many locusts are there that weigh 0.01 g? The same argument holds true for *extrapolation*, the process of extending the theoretical relationship beyond the range of the data set. Do you know of any locusts that weigh 5 g? Apart from the fact that they would be big and scary, can you be sure that they could jump 2 m? The rule of thumb is that if you draw a straight line through the data set, it should only extend over the range of the data points and not beyond unless you have strong evidence that the relationship will hold true for other data points.

9 Researching your coursework topic

→ Introduction

Whether you are looking to read around a topic to improve your understanding of the material covered in your lectures and tutorials or preparing for an assignment, you will need to undertake research.

In the previous chapters, we have discussed the process of reviewing assignment briefs, so you know what is expected of you, and thought about the process of structuring assignments. Core to preparing your assignment submissions is the incorporation of taught content from the course along with material from your own research, with the latter becoming increasingly important as you progress through your studies. It is also essential, as you include the work of other people or indeed use AI to assist you, that you adhere to the principles of academic integrity and avoid plagiarism.

Researching the topic is probably the most time-consuming part of most assignments. Your lecturers will no doubt hope that you are really enthused by the research you undertake and that you will explore the subject in depth. That's great, but you need a good dose of pragmatism in your approach and to be aware that undertaking the research could be literally an endless task. Therefore, you need to keep your timeline in mind and be strategic in your approach.

9.1 What is the purpose of researching a topic?

Independent learning is an essential aspect of most bioscience degrees, arguably more so than any of the other science subjects. From the earliest stages of the degree, Bioscience students are expected to expand their knowledge beyond what is taught in lectures. This is particularly evident in the essay format of most biology exams and the comparatively large amount of coursework compared to the rest of STEM subjects. In order to do well on these types of assignment you will need to consult scientific sources you have found yourself that expand beyond what you have been taught directly. We will now take a little bit of time to explore why this is such an important aspect of Bioscience degrees, so that you can understand exactly what it is your lecturers are expecting you do to.

There is also significant value in becoming proficient at researching topics because the skills required to effectively learn a subject through independent research/study are highly transferable to your personal and professional futures. The future is uncertain (technology disrupts), knowledge changes, and university cannot prepare you for every exact profession you might enter after you graduate. The capability to rapidly 'become expert' as you enter new intellectual environments is highly beneficial. Similarly, the ability to know where to find information sources, the ability to understand them, critically evaluate them, and synthesise them into an informed and evidence-based opinion is something most professionals do regularly and the foundation of effective citizenship.

What does researching the subject for coursework consist of?

- Finding and understanding multiple sources of complex scientific knowledge that will help you understand the subject you are communicating.

- Sorting between appropriate and inappropriate sources of information to evidence the points you make in your assignment.

- Critically reviewing multiple perspectives on a topic and synthesising them into a coherent narrative that answers the question and/or assignment brief.

Ultimately, to research the subject is to build on the work of other people. To understand what has been done before, to bring together different ideas on a subject, to understand how your results fit within the context of known science, and communicate that to others.

9.2 The importance of academic integrity

9.2.1 What is academic integrity?

Academic integrity underpins all academic endeavour because it reflects the trust that can be placed in the work that is being presented. The International Centre for Academic Integrity (ICAI, 2021) identified six Fundamental Values of Academic Integrity:

- Honesty
- Trust
- Fairness
- Respect
- Responsibility
- Courage

The term 'academic integrity' was developed by the American academic Don McCabe. In its broadest terms, it means being open and honest in all your academic work. In particular, this means giving due credit to those people whose ideas or work you use in your own work and being honest in presenting the outcomes of your research—for example, not falsifying results. This is essential if your work is to be trusted and assessed fairly.

Academic integrity means that whenever you make use of the ideas, writings, images, or other materials that have been created by other people, you should acknowledge that and not give the impression that these were your own. Not to give that acknowledgement is plagiarism, which is a form of academic misconduct and can have serious negative repercussions for your studies (plagiarism and academic misconduct are covered in more detail in sections 9.5 and 9.6). Presenting written, visual, and oral content created by other people (e.g. contract essay mills) or other 'intelligences' (e.g. Generative AI) as your own is another form of serious academic misconduct, which is basically the opposite of academic integrity.

9.2.2 Why academic integrity?

Why are we so concerned about academic integrity? Why does it matter so much if I submit work completed by somebody [or something] else as my own? Are they overreacting or are there good reasons for their concern? Well, let's think for a minute about what underpins academic work. To be more specific, let's consider what is good academic practice.

Academic integrity underpins all of science!

It was Isaac Newton who said of himself (in a letter to Robert Hooke in February 1676):

> If I have seen further, it is by standing on the shoulders of giants.

By 'have seen further' he was referring to the tremendous discoveries he had made during his lifetime; by 'standing on the shoulders of giants' he was largely referring to the people who had gone before him, whose ideas he developed and built upon. In making this statement Newton was highlighting a very important principle of academic study: that although you might research and then write up something individually, you are, in fact, contributing to a collective endeavour. Even the most novel or original of discoveries will build upon the ideas of others, and it is important that those ideas are acknowledged and that it is only your own contribution that you claim for yourself.

So, academic integrity is important because it underpins good academic practice: it is fundamental to all of scientific knowledge.

Academic integrity ensures that you learn what you are supposed to learn

When you are set an assessed piece of work, it is easy to focus on the tangible endpoint, that is, the mark you are awarded. However, you are set work not merely as a means of allocating marks, but also as a means of helping you to learn. Finding, understanding, evaluating, analysing, synthesising, and communicating information are all fundamental competencies you will need for the rest of your personal and professional lives. When you plagiarise, you do not actually do any of the work, which means you won't learn how to actually apply any of these essential skills within or beyond your course.

Academic integrity protects you and the wider community

Completing your degree is certification that you have developed your knowledge, skills, and attributes. As outlined above, when you plagiarise, you short circuit the learning process and do not actually develop the things your degree certificate says you have. Imagine if this applied to your doctor, would you still feel like they were qualified to do their job? Similarly, if you are employed when you graduate, your employer will expect you to be able to do the things your degree says you can do. If you continue to plagiarise at this stage, the ramifications would be much more serious than a penalty to your grade.

Academic integrity helps you build your own perspective

Many people consider independent thinking to be one of the most important skills developed at university. Developing your own conclusions on complex topics that are supported by appropriate evidence is an important academic and personal milestone. It applies equally well to complex scientific subjects as it does to sociopolitical ones. If you plagiarise by not commenting on or analysing evidence, you are not thinking for yourself; you are merely copying the ideas of others and have failed to develop your own point of view.

Academic integrity underpins academic fairness

How would you feel if a student on your course scored a higher mark than you on an essay, but you knew that they had cheated? Make sure you don't make anyone else feel the same way about you.

Academic misconduct can result in severe penalties

Different universities have different penalties for students who have plagiarised. The type of penalty depends on the precise nature of the plagiarism, the extent to which the work contributes to final marks, and whether or not it is a first offence. Common penalties include:

- an awarding of zero for the piece of work, the requirement to repeat the work, plus an official warning;
- an awarding of zero for the module and withdrawal of the right to resit;
- the downgrading of degree class by one division;
- expulsion from the course.

Many universities are now using plagiarism detection software. Students are required to submit their work electronically, the software then compares the work with a vast array of other sources, including core texts, journal articles, internet sites, and previously submitted student work. A report is produced highlighting the areas of similarity with the compared sources. The tutor would then interpret the report to decide whether or not it is plagiarised. The software is becoming increasingly sophisticated and makes getting away with plagiarism much more difficult. However, even if your university doesn't currently operate an electronic detection system, it is worth noting that academics are also very capable of quickly spotting plagiarism using more old-fashioned methods.

It is also worth noting that words, ideas, data, and images that have been written, thought of, discovered, or created by someone else belong to that person. Using them without acknowledging the source is stealing. You could be in breach of intellectual property or copyright laws, which would be incredibly serious, especially in the 'real world' outside of university.

> **BOX 9.1 Learning not copying**
>
> Remember that avoiding plagiarising is not simply about evading punishment and the purpose of doing a degree is not to obtain a piece of paper. You do the degree because you will learn things that are important for your future and your career. It may be possible to pass a module by getting a Generative AI, an essay mill, or someone else to create your assignments for you, but you will not actually have learned anything. This could get you in serious trouble during your studies or later in your life.

9.3 Finding appropriate sources

Hopefully by this stage you understand why the research process is important, what you are supposed to achieve, and why it is important that you approach it with academic integrity; but what is the best way to go about actually doing it?

9.3.1 Use appropriate sources

There is an increasing amount of material available for research purposes, and you need to be strategic and critical in the way in which you use these different resources. There is some additional guidance on critical approaches to writing in Chapter 12, which you may wish to consider once you are more familiar with the contents of this section. The main sources of information are listed, and it is probably a good strategy to work through them in this order so as to keep a clear focus. This approach also means that you are starting with relatively simple, well-defined information and progressively moving to the more complex, wider sources.

We have discussed details of many of the key sources of printed information in Chapter 4, section 4.2 *Making the most of the library*. So, we will only summarise these here:

- *Lecture notes and handouts*: these will often provide the starting point for your investigations, by reading through what your lecturers have already told you about the topic.

- *Textbooks*: If you are in the very early stages of your university career, textbooks may provide the bulk of the information you need to develop your understanding or to undertake your assignment. However, as with any source of information, textbooks do have limitations in terms of the depth of coverage and currency.

- *Reading lists*: for many courses you will be given lists of suggested reading that provide additional subject depth and breadth, to enable you to develop your understanding of the topics being covered. These lists will normally comprise journal articles and similar materials that have been selected carefully by your lecturers as being appropriate for the course you are studying. You may even be given some suggested titles as a starting point for your assignment.

- *Specialist topic books/monographs*: As their name suggests, these are books that focus on a specific topic within the subject and can provide you with a great deal of focused information setting out overviews of current research. As with textbooks, though, they tend to go out of date quite quickly.

- *General science journals*: There are several general science journals, such as *New Scientist* and *Scientific American*, which publish review-type articles aimed at the general interest reader with a scientific background. These types of journal can be very useful in providing overviews of a specific topic.

- *Specialist review articles*: Many research journals include review articles as well as the papers describing the results of novel research. There are also numerous journals dedicated to reviews in the biosciences, such as the 'Annual Reviews in . . .' series or the 'Trends in . . .' series. Review articles often make very useful starting points in research for coursework assignments, both because of the overview they provide and because of the references made to recent research work in the field.

- *Research papers*: as you progress through your course, there will normally be an increasing expectation that you will be reading and using research papers that you have found in research journals in order to develop the subject content of your assignment.

As with the review articles, your lecturers will certainly include recommended research papers (some of which they have probably written themselves!) in the reading lists they give you and may even suggest some articles as starting points for your assignment.

As you progress through your studies you will be increasingly expected to make use of the final two sources of information in the list above (i.e. Review Articles and Research Papers) in your assignments. It is entirely possible that, by your final year, the entire contents of an assignment are drawn from reading that you have done yourself with limited guidance that extend out beyond anything you have been taught during the lectures on your course.

9.3.2 Finding research papers

Through your university library you will probably have access to useful search engines, such as Ovid, Web of Science, PubMed, and Medline, that you can use specifically for searching through the vast number of research journals that are available in electronic format. You can also use Google Scholar, which is available to everyone.

Different databases cover different ranges of journals, and so they vary in size and in terms of the number of journals included. Some, such as Web of Science, are relatively broad; others, such as Medline, are more specific, only including journals likely to be of relevance to medical research. You will need to select which database you use and should consider seeking advice from a university librarian, who will have a good knowledge of the databases your university subscribes to and the range of topics that they cover.

Your librarian is also likely to be able to give you some initial advice about how to search for research papers. You will need to be careful in developing your search pattern. You can do this by selecting the appropriate keywords and seeing how many hits you obtain.

Gradually refine your keyword selection until you have a limited number of hits that should be closely relevant to the subject matter of your assignment.

As an example, let's consider the following assignment:

> Discuss the different biochemical, physiological, and neural mechanisms contributing to muscle fatigue in humans.

So, you could start with a crude search in Web of Science and simply enter the words 'muscle fatigue'. This comes up with almost 32,000 research papers. Similarly, if the same keywords are entered into Google Scholar, more than 750,000 articles are listed! In either case, you definitely don't have time to scan the titles, let alone read the articles! So, how can you refine the search strategy to pull out the articles that will be of value?

There are several basic options here:

- choose specific types of article, for example review articles or research papers;
- select only those papers published recently, for example the last five years;
- refine your search terms.

Choose the type of article

The search engines allow you to specify the type of article you would like to select. So, you could search specifically for review articles, or you can exclude meeting abstracts (because they are likely to be short and very narrow in focus). If we specify only review articles, then Web of Science returns 2,650 and Google Scholar 340,000. Still far too many but significantly fewer compared with our starting point.

Select only recent publications

An initial selection by date of publication can rapidly reduce the number of hits. This allows you, for example, to narrow down to a single year if you wish. A rule of thumb for your assignment might be to narrow down to the last five years as your starting point. For example, with Google Scholar, limiting to review articles published in the last five years reduces the number of hits to 22,600 and Web of Science returns 886.

These are still too many to work with, but you will have certainly reduced the field. There is a risk, however, that you may exclude some older but very useful articles.

Refine your search terms

The next option is to refine your search terms, after all 'muscle fatigue' is a very broad term to use for searching.

The title of the assignment actually referred to the 'mechanisms of muscle fatigue'. So, a further refinement would be to specify 'mechanisms of muscle fatigue', so you could use that specific phrase for your search, which will greatly reduce the number of articles,

Figure 9.1 First page of a PubMed search on the topic of 'mechanisms of muscle fatigue'. The initial selection was for review articles from the last five years and it returned 185 results.

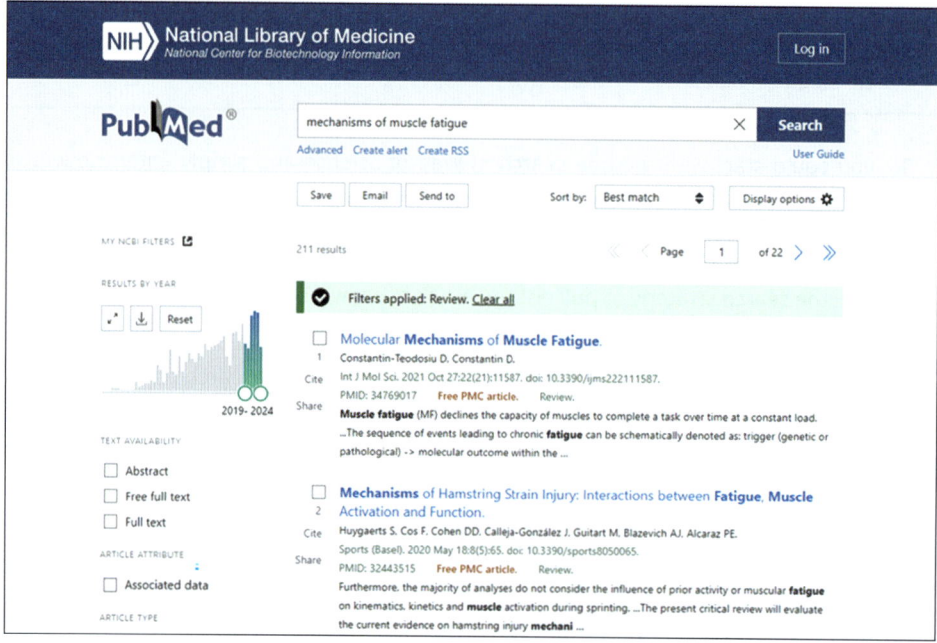

especially if you also only search within the last five years. Web of Science now returns 215 review articles published in the last five years. Using a similar selection process, PubMed returns 185 articles (Figure 9.1):

This was a fairly simple strategy, but you may find that not all searches are as simple and that you need to be more sophisticated.

Some search engines have inbuilt facilities to add search terms sequentially to refine your search. For example, if you specifically want to find out about muscle fatigue research that has been done on humans you could add 'human' as part of your search string:

> muscle fatigue AND human

This will now return a list of articles where the words muscle and fatigue occur together and the word human also occurs. However, if you wanted to find out about any work that had been carried out on animals other than humans, you would enter the search string:

> muscle fatigue NOT human

It may also be the case that you want to broaden your search strategy to include related terms, particularly if your initial search has only returned a small number of articles. You could think of all the variations of the key term and type each one into your search phrase, which is not a very efficient approach, or you could use a truncation symbol such as *. If you enter the search string:

```
musc* fatigue
```

this would return all the terms with 'musc' as the base, such as muscle and muscular. This can be very useful, but again there is a risk of getting too many responses, not all of which will be relevant.

Each time you undertake a search, you should look at the number of articles that the search has returned and scan the first few titles to check that they appear to be relevant. With a broad search you are likely to find a large number of articles, many of which are not directly relevant. As you refine your search, for example, by selecting specific years or by using more specific search terms, then the number of articles should progressively reduce and should become more relevant, but there is a compromise as you may also start to exclude some articles that would be useful.

Coming further up to date

In your reading list, or from your searches, you may have an article that is very interesting, but is four or five years old. So, you want to find something more recent that is directly related. If you have obtained the details of the article from a search engine, you can use that article as a starting point and search using citations. In other words, the database search looks for more recent articles that cite the one you have obtained. This enables you to bring your research up to date, following the specific theme of your original article.

After you have completed your searching, you will probably still have several tens of articles in your list and reading all those in full would, clearly, be very time-consuming. Reading the titles of the papers your search has come up with is therefore the next stage of selection. When you have drawn up a manageable shortlist of the titles that appear most relevant, then you can go to the next stage of reading the abstracts and, from your final list, select those papers which you need to read in full (having first checked that you can get access to the full articles!).

Try this—Finding different types of article

Think of a topic that you have just covered in your course and set yourself this test:

- Find five review articles on the topic that were published in the last three years.
- Then, find five articles published within the last three years that address a specific aspect of the topic.
- In doing this exercise, experiment with the search terms using some of the processes covered in the preceding section.

Try the exercise using two different databases and see if you end up with the same articles!

9.3.3 Web resources

The major web search engines, such as Google, Yahoo!, Bing, etc., are familiar to most people and they can be used to scan literally billions of pages of web-based information. They are very effective at producing long lists of websites related to your inquiry, but this means you have to be very selective in terms both of finding material that is directly relevant and then evaluating the reliability of that material. Don't forget that anyone can host a site and put information onto the web, and there is no guarantee that the information has been checked for accuracy. The material on the web therefore varies in quality from the excellent to the downright wrong.

If we try out our search for muscle fatigue using general search engines, Google, for example, will return about 80,000,000 results! The ranking of the list, however, does not necessarily indicate the direct relevance of the information to be found on the site (see Box 9.2 regarding rankings). You will also see a lot of different adverts on the pages, for example, for dietary supplements to help improve your muscle strength!

If the title is on a subject of more general interest, even more results are very generated. For example, consider the 'effects of vaping' as a title. In this case, Google returned 107,000,000 results.

These links will be to a whole variety of sites, such as commercial sites offering products to help you stop smoking, government advisory sites, pressure groups (from various perspectives), media sites, such as the BBC, charities, and self-help organisations, medical advisory groups such as NHS Direct, and many more. Therefore, you will have to exercise judgement in terms of the types of sites you look at and you need to determine their reliability.

How reliable is a website?

As we have seen, there are many sites providing information on specific topics, and some are more reliable than others. So, how do you know what to use? From a purely academic standpoint, the scientific journals, with their peer-reviewed papers, offer the most reliable sources of information, but there are many other useful sites as well. The simplest questions to ask yourself about each site are: who or what organisation wrote the material, and how old is it?

BOX 9.2 Web search engines: a very brief introduction

When we are looking for information about muscle fatigue and enter those words into the search box of a search engine, the search engine will return sites that contain the words 'muscle fatigue', but the words will not necessarily be together, or in that order, unless you have used an advanced search strategy. The order of the listing of the sites is also determined by a ranking process. This process ranks pages on a number of criteria, one of the most important of which is the number of 'backlinks'. These are the links, for example, hyperlinks in text, feeding to this page from other websites, so the more links there are to a given web page and the higher the ranking of each of those sites, then the higher that web page will appear on the search engine's listing. The use of specific keywords also contributes to the ranking algorithm as does optimisation for mobile devices.

You can quickly answer these questions by looking at the site address in the listing. Each site is identifiable by its universal resource locator (URL). Examples of useful sites would include:

- *University sites*: these can include course materials from the university's teaching programmes, which can be very useful as an additional resource, as well as providing links to the publications and research outputs of their academic departments. Sites for higher education institutes are identifiable from the URL, which will contain the code ac (academic) or edu (educational). Thus, universities in the UK all have an address that takes the form: www.[universityname].ac.uk. Those in the USA have the address www.[universityname].edu, and for Australia they are www.[universityname].edu.au. Many European universities also have extensive websites in English and are useful sources, so, for example, French university addresses take the form of www.[universityname].fr/en

- *Government sites*: these can provide information at a range of levels, particularly in terms of policy or public advice. Examples here would be the Food Standards Agency, the Environment Agency, or the National Health Service (NHS). Most government sites in the UK have the address form that begins with www.gov.uk followed by the department title, or www.[title].gov.uk. For example, the link to the Environment Agency is https://www.gov.uk/government/organisations/environment-agency. NHS sites similarly have the address form: www.nhs.uk/[name] or www.[name].nhs.uk.

- *Learned societies*: there is a learned society associated with almost every specific area within the Biosciences; for example, the Society for General Microbiology, Physiological Society, British Neuroscience Association, Society for Experimental Biology, and many more. Most of these sites have addresses in the form: www.[name].org. Addresses ending in .org are used for organisations that are non-governmental and non-profit-making.

- *Professional bodies*: these are organisations that represent and maintain the professional standing of specific professions, such as the General Medical Council, the General Dental Council, and the Royal Society of Biology. These sites, too, usually have addresses as: www.[name].org.

- *Media organisations*: many of these organisations, such as the BBC, The Times, The Guardian, and The Daily Telegraph, are very useful for providing information about topical science stories. Remember, however, that the story is usually written up by a journalist who may have a limited knowledge of the topic and is writing for the general public, rather than for someone looking for detailed, scientific information. Media organisations are also in the business of selling news stories and, therefore, they may focus on the more sensational aspects of a topic. The BBC can be found at https://www.bbc.co.uk/news. In the case of newspapers, as these are commercial organisations, their URL will end in .com.

- *Commercial organisations*: many commercial organisations have very useful websites, but it is important to remember that they also want to sell a product. These may carry quite a lot of useful educational information, however, they also carry a lot of commercial information setting out the benefits of their specific product. Although the information published by such sites has to comply with the requirements of the Advertising

Standards Agency, they do not have to be objective in their approach. For example, while they are extolling the benefits of their own product, they are under no obligation to inform you that another company's product may actually be more effective.

- *Specific societies and organisations*: there is a huge variety of organisations and groups with websites, and the quality of these varies greatly. In the area of health, there are numerous self-help organisations that have sites that carry useful information, such as the Parkinson's Disease Society, the Alzheimer's Society, or the Motor Neuron Disease Association. In environmental biology, there are organisations such as the World Wildlife Fund and Greenpeace. These sites are typically aimed at a non-academic readership, but often carry a significant amount of background information. However, these organisations frequently have very focused agendas and, therefore, the information provided may be selective or interpreted in very specific ways, so, when you are reading, you need to be aware of that focus and to maintain your critical approach. Most of the addresses for these types of organisation will end in .org.

When you are looking through the journals of professional scientific bodies, you have the confidence of knowing that the research papers have been reviewed by scientists who know the subject area very well. Therefore, you can be confident about the validity of the material being presented. By contrast, the material that is posted on websites may have been posted by experts in the field, but it may also have been put there by people who have little or no real knowledge, but who have views about a subject that they want to broadcast to the world.

So, you need to be cautious and critical, and look for specific signposts that can indicate the reliability of the information presented: for example, if the website is published by a university or a professional body, then the information is likely to be accurate and reliable. If it is published by a commercial organisation, for example, a drug company, it is likely to be accurate, too, but may also be subject to a certain amount of marketing 'spin', for example, to encourage take-up of a specific drug. At the other end of the spectrum, if the material is published by an individual or organisation with no obvious professional qualifications or academic affiliations, then you should treat the site with considerable caution.

Wikipedia

We also need to consider Wikipedia. Wikipedia is often the first point of a search to give a perspective on a topic. Wikipedia is a freely available online encyclopaedia to which anyone can contribute. In this way, it differs from other encyclopaedias where individual sections are usually written by experts in the field. Wikipedia is continually increasing in volume, but because of its nature, the quality of the material is variable: much of the material is of high quality and has been written by experts in their respective fields. Some, however, is less ac-curate and is not necessarily reliable. Unless you are an expert who already knows the field, the problem you will have is picking out what is good from what is not. Wikipedia can be useful, for example, for reminding yourself quickly about a topic or as a starting point for some research, but it is not necessarily advisable to use it if you are learning about a topic from scratch or researching for an essay.

9.3.4 Research strategies for better assignments

It's past midnight and you are finally in the process of writing your essay. It's due in for tomorrow morning, but that's okay because you have done all the research, you have your plan in front of you and the writing is going well. Then you hit a problem—you just need to describe the key experiment that supported the argument you are putting forward. You know you read up on it, but you can't quite remember the details. It's somewhere in that pile of review articles, but you can't remember which one and you don't have any notes to jog your memory . . .

Situations like this are very frustrating, but they can easily be avoided if you are systematic in making notes when you read up on a topic. Many students only think of note-making in the context of lectures, but it is just as important in terms of your reading around the subject, for example, when preparing for an assignment.

9.4 What should you include in your notes?

Good note-making is an important aspect of academic integrity and will ensure you know exactly who to credit, so you do not accidentally portray other people's work as your own (and be accused of academic misconduct by your lecturers).

To ensure that the notes from your reading are as useful as possible and that your note-making is done effectively, you must always record the source. When you come to include the information you have read, it is essential that you can include the full citation from the reference (Figure 9.2): just jotting down the title of a book is of limited help, particularly if it is several hundred pages long! Don't forget also that if you are note-making for an assignment, you will need the full reference for your reference list.

As part of the preparation for your assignment, you should have analysed the question (Chapter 7, section 7.3) and identified some key points you need to write about (Chapter 8). We highly recommend that you align the note-making from your research with your assignment planning. Research should help you understand what to include to meet the assignment aims, and your knowledge of what you need to achieve should inform your research strategy. There's no need to make notes detailing the whole contents of a source if the content you need to refer to is covered in a small section.

Figure 9.2 Reference details for notes from a book and a research paper. It's okay to use abbreviations in the details since you can go back to the full reference if you need it for your assignment.

Scott, J. et al (2023) Biological Science: Exploring the Science of Life Oxford Univ. Press. Chap 20, p696

Wan, J. J. et al (2017) Muscle fatigue: general understanding and treatment. Experimental & Molecular Medicine 49(10) e384.

We consider three specific examples of this in section 8.1.4, where sources may be used to:

1. illustrate a point (as an example);
2. provide evidence to support a point;
3. provide evidence of a contrasting argument.

When making notes, try to think about how that source helps you create a narrative that meets the assignment brief. Thinking about what point you need to make that the study supports, contradicts, or exemplifies. Thinking about your research notes is helpful because:

- It encourages you to incorporate the ideas into your assignment using your own words (which helps to avoid plagiarism).
- It will help you write in a focused way. Making sure that all the points you make are relevant, which allows you to maximise the content within the word count.

Turning notes into writing in Chapter 8, section 8.1.4 provides additional guidance on how to turn notes into coursework but you can lay the groundwork for this by ensuring that you remember what the aim of the assignment is when you make your notes. Chapter 12, section 12.2.3 contains guidance on how to rapidly extract key information from scientific papers that are reporting on the findings of a specific study.

The quality of the evidence

This is dealt with in detail in Chapter 12 on critical thinking, however, we will touch on it here. As you progress through your studies you will be increasingly expected to not just read the scientific literature but critically reflect on it as well. Particularly, when you present a point, position, or fact within your assignment, you need to qualify whether it is a well-supported point. In order to do this, you should make some notes on what evidence the study has to support the point being made. How have the data been collected? Are there any limitations in the method that limit the scope of the conclusions?

If you are at a stage of your studies where you need to be including critical thinking in your assignments (check the marking criteria), then go to Chapter 12, section 12.2.5 for a method to embed critical thinking into your research process.

9.4.1 A strategic approach for researching an assignment —the inquisitive approach

Researching your topic can be an overwhelming task. As we have seen above, even a highly targeted search strategy on mechanisms of muscle fatigue returns ~185 relevant research papers. This is probably more research than you are likely to be able to review within a sensible timeframe, even if you are a final year student working on your dissertation literature review. So, can you do anything to make this stage of the process more manageable?

It can help to think of the research literature as a bank of answers to scientific questions. You can make your research process less overwhelming by knowing exactly what question(s) you want the answer to. You can help do this by formatting your assignment plan a very specific way.

Let's take the assignment title above:

> Discuss the different biochemical, physiological, and neural mechanisms contributing to muscle fatigue in humans.

Now, instead of a general search on the topic of muscle fatigue, you could structure your plan into a series of very specific questions that build up to provide an answer to the overall question posed in the assignment brief. You can start off with some broader generic questions, then, as you learn more about the subject, you can ask more detailed questions of the research literature.

Example plan

Introduction

Question A: What is muscle fatigue (in humans)?
Answer: A definition taken from a textbook, lecture, or even a paper on the subject.

Question B: What are the different mechanisms of muscle fatigue?
Answer: You might find a review paper that covers the state of knowledge of the different mechanisms in 2015, it is structured into two main sections: biomechanical versus biochemical mechanisms. It's an easy paper to read, so you decide you want to use the same structure for your assignment.
   Assignment Section 1 (Biochemical Mechanisms) Paragraph 1 (Introduction to Biochemical Mechanisms)

Question C: What are the biochemical mechanisms of muscle fatigue?
Answer: You might find a paper that outlines four biochemical pathways that you need to cover: lactate build up, ion concentration related to action potentials, calcium ion concentrations, and reactive oxygen species. So, you decide to do a paragraph for each of these in your assignment.
Assignment Section 1 (Biochemical Mechanisms) Paragraph 2 (Lactate build-up)

Question D: How does lactate determine muscle fatigue?
Answer: You might find a paper suggesting that decreasing pH (from increased lactic acid) reduces muscle force production.

Question E: Is there good evidence that lactate determines muscle fatigue?
Answer: However, a more recent paper might point out that more recent studies suggest a role for other mechanisms.
   You can repeat the questions D and E for each of the different biochemical mechanisms, and then repeat the whole process for the biomechanical mechanisms.

Finally, you can draw all the different pieces of evidence together in a concluding paragraph to summarise your answer.

> Discuss the different biochemical, physiological, and neural mechanisms contributing to muscle fatigue in humans.

What this means is that every time you dive into the literature you only need to find the answer to a specific question as opposed to just improving your understanding of a broad topic.

You can combine this technique with the guidance on how to structure paragraphs in section 8.1.4 to help work out what questions you need to answer. Once you have the answers to each of the sections of a paragraph it should be relatively simple to turn the plan into a first draft.

You can use this method to improve your critical thinking as well, by asking questions such as:

- Is there good evidence to support the answer above?
- Does anybody disagree with the answer above?
- What are the limitations of the methods used to reach the answer above?
- What research could be done to provide us better answers to the question above?

We will deal with this more in the section on critical thinking (Chapter 12, section 12.2).

9.4.2 Referencing and referencing software

As you assemble your notes and the sources of the material, you will be building up a list of different references that you need to cite and include in your bibliography. If you are engaged in an extended piece of writing, then this list can quickly become difficult to manage and format correctly. This is where referencing software can be very useful, especially as you can use it to compile lists for more than one piece of work. These programs commonly integrate with standard word-processing software, such as Word™.

Many university libraries subscribe to one or more reference management programs which are available to students and staff. Examples of these programs in widespread use include:

- EndNote
- RefWorks
- Mendeley

Many modern versions of these referencing software allow you to attach pdf copies of the paper and even help structure your note-making process. If you expect to have a lot of coursework on your course, it can be worth investing time in learning how this software works early on.

9.5 Academic misconduct (deliberate plagiarism)

Near the beginning of this chapter, we introduced the concept of academic integrity and its opposite academic misconduct. When you are writing or presenting any piece of work, it is very important that you acknowledge the sources of the information and ideas that you are presenting. If you present an idea or a piece of information without acknowledging a specific source, then you are effectively claiming ownership of that idea or information. Provided it is your idea then that's okay: for example, it could be your interpretation of the results obtained in a practical class or from your research project. The failure to acknowledge published sources, however, is one of the most common reasons for being accused of plagiarism.

In the context of plagiarism, 'academic misconduct' is any deliberate attempt to pass off the work of other people (or intelligences) as your own. 'Academic poor practice' is when this is done by accident, often through insufficiently rigorous notetaking, referencing, or incorporation of ideas into your own unique narrative. Both can result in serious penalties on your studies and we have detailed extensively already why academic integrity is an essential part of the scientific and educational process (section 9.2). The rest of this chapter will attempt to explain different forms of plagiarism to make sure that you understand what they are and how you can avoid them through good academic practice.

9.5.1 Deliberate plagiarism (copying)

The Oxford English Dictionary defines plagiarism as:

> The action or practice of taking someone else's work, idea, etc., and passing it off as one's own; literary theft.

So, plagiarism is about using someone else's work but giving the impression that it is your own. Let's think for a moment about the different elements of this definition:

- *taking*—could be from a book, journal, internet site, a lecture or handout, or even another student's work;
- *someone else's work*—could be someone else's words but could also be their ideas, data, or images;
- *passing it off as one's own*—could be either deliberate or accidental and could be in the context of an essay, a lab report, or a presentation (in fact, any assessed piece of work);
- *literary theft*—tells you in unambiguous terms how it is perceived by many, including your tutors.

So, put simply, plagiarism is a form of cheating. You wouldn't cheat in exams, and you shouldn't cheat in coursework either. Therefore, just as there are penalties for cheating in exams, there are penalties for cheating in coursework too.

When people hear the word 'plagiarism', they probably think of someone deliberately copying text without referencing it but, as we have seen from the definition, plagiarism is often much broader than simply copying text. We will deal first with examples of deliberate plagiarism.

Imagine that a student used the following text in an essay on 'Factors influencing the development of cancer'.

> There are many factors which influence the development of cancer. These include both endogenous factors, such as inherited predisposition, and exogenous factors, such as exposure to environmental carcinogens and infectious agents. Another factor which has a clear influence on the type of cancer which develops is age.

Figure 9.3 Sample of a plagiarised image. Image taken from Knowles and Selby (2005), Figure 1.3, p. 8

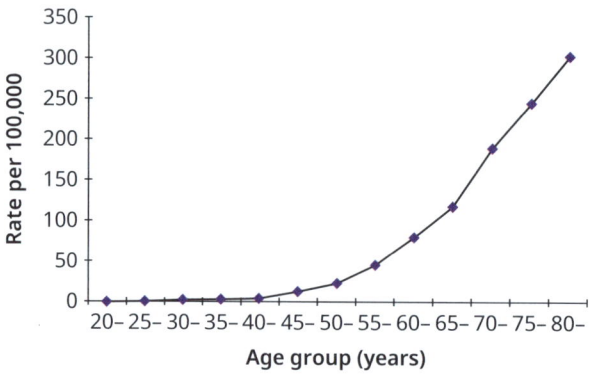

The first sentence is the student's own but the next two sentences are not written by the student; rather they are taken from the *Introduction to the Cellular and Molecular Biology of Cancer* (4th Ed) by M. A. Knowles and P. J. Selby (2005). The student has copied the text without any indication that some of the words have come from another source. They have therefore plagiarised by copying.

Alternatively, take an example of an image: imagine a student had used Figure 9.3 in an essay on 'Factors influencing the development of cancer'. If the student failed to acknowledge the source of the image, in this case Knowles and Selby (2005), Figure 1.3, p. 8, they would also be plagiarising by copying. It is important to remember that plagiarising by copying images is equivalent to, and so just as serious as, plagiarising by copying of text.

9.5.2 Collaboration and collusion

The boundary between collaboration and collusion is one that students often find confusing. At one end of the spectrum, you may be required to collaborate in group work exercises in which a single piece of work is produced representing the combined efforts of the group. Often such pieces of work may be awarded a single mark that is given to all the members of the group. At the other end of the spectrum, two students may work together on a piece of work that is supposed to be an individual exercise. The latter is an example of collusion since the assignment is supposed to reflect the work of the individual students working independently. Under those circumstances, the students would normally be subject to the same penalties as for plagiarism.

You should also be aware that, although you have undertaken the work individually, if you knowingly allow another student to copy your work, you will both be in breach of your university's regulations. In this case, remember that this also applies if the student is in a different year, for example the year below you, or is studying at another university. Because plagiarism checking systems such as Turnitin™ keep a database of all submitted work, there is automatic cross-checking between year groups and institutions.

Figure 9.4 Collaboration in lab write-ups

Student A	Student B
Discussion	**Discussion**
In these experiments, observations were made of the changing patterns of electrical activity in skeletal muscles during voluntary contractions. During contractions of progressively increasing force, there was an increase in the electrical activity recorded from the muscle. This increase was related to the two different ways in which the nervous system controls muscle force: an increase in the action potential frequency of individual motor units and the recruitment of additional motor units (Berne and Levy, 2000).	Recordings have been made of the patterns of neural activity in skeletal muscles during the course of voluntary contractions. When the force was progressively increased, there was an increase in the electrical activity recorded from the muscle. This increase was attributed to the two ways the nervous system regulates muscle force, namely by increasing the action potential frequency of individual motor units and by recruiting additional motor units (Berne and Levy, 2000).
Observations of single unit activity showed that the firing began at about 8 impulses/s, increasing to 40 impulses/s at maximum force...	Recordings of the activity in single motor units showed that the discharge began at about 8 impulses/s and increased to 40 impulses/s at the maximum force...

The boundary becomes more blurred where part of the activity requires collaboration, but the final product has to be an individual effort. The typical example of this would be working with a partner in a practical class: it is important to remember that whilst you will often need to work with a fellow student to conduct a practical experiment or undertake fieldwork, this doesn't usually mean that you are being asked to collaborate on the writing-up aspect of the report as well. Obviously, there will be similarities between two reports written by two students who had worked together to conduct an experiment, but these will largely be confined to the methods and results sections. However, in the analysis of the results and in the discussion section, you will be expected to analyse the data and draw conclusions from the findings independently of each other. For example, compare the two discussion sections in Figure 9.4 from students A and B who were lab partners for a practical class on electrical activity in skeletal muscles.

You can see that there are definite similarities between the two discussions; the marker would probably conclude that the students had collaborated on the write-up of the discussion section and had, therefore, plagiarised.

Now look at the examples of students A and C in Figure 9.5.

Although we would still expect to see similarities in the methods and results sections, we can see that the students have thought independently about their findings and have come to their own conclusions. This doesn't mean (in this case) that they have come to different conclusions, but the wording and sentence structure suggests that they have thought the ideas through for themselves.

Figure 9.5 Independent lab write-ups

Student A	Student C
Discussion	**Discussion**
In these experiments, observations were made of the changing patterns of electrical activity in skeletal muscles during voluntary contractions. During contractions of progressively increasing force, there was an increase in the electrical activity recorded from the muscle. This increase was related to the two different ways in which the nervous system controls muscle force: an increase in the action potential frequency of individual motor units and the recruitment of additional motor units (Berne and Levy, 2000).	The aim of the experiments was to use electromyographic (EMG) recordings from the whole hypothenar and from single motor units to investigate the neural control of muscle force during voluntary contractions. The initial test contraction involved a controlled increase in force and the recordings showed a progressive increase in the level of EMG activity paralleling that of the force production. It was predicted that this increase in EMG activity was made up of two different elements of neural control: variation in the rate of action potential discharge in the individual motor units and recruitment of additional motor units (Berne and Levy, 2000), these two processes occurring in parallel.
Observations of single unit activity showed that the firing began at about 8 impulses/s, increasing to 40 impulses/s at maximum force…	To test this prediction, recordings were made of the activity of single motor units . . .

9.5.3 Self-plagiarism

It is also possible to self-plagiarise. This is when students submit their own work, which they have previously submitted for another part of their course but fail to declare that it had been used for such a purpose. This is probably the type of deliberate plagiarism that students find the most frustrating; 'I'm the author of the work', you might reason, 'so surely I can submit it as my own work for whatever aspect of my course I choose'. It can be difficult not to have some sympathy with this view, but nevertheless self-plagiarism is considered unethical because of the deception involved in submitting the same material for credit in different modules or courses. On a more positive note, writing can always be improved upon, so it is in your own interest to do more than just recycle previously submitted work; it also helps the learning process. This is in addition to the more obvious point that a good answer written for one essay question, for example, is probably not going to be specific enough to be considered a good answer to a different, even if related, essay question. In short, although disallowing self-plagiarism may seem a little harsh, it is to your advantage to tackle each piece of assessed work as a separate exercise; allowing you to tailor your answer to the specific task and broaden your knowledge of the subject.

9.5.4 Buying

There are numerous 'essay mills' through which it is possible to buy a range of pieces of work, including essays, dissertations, etc. and which typically claim to be checked to ensure they are plagiarism free. Since 2022, essay mills have been illegal in the UK but it is still very easy to access their services. This is one of the most blatant and outrageous forms of plagiarism there is. We have already explained why academic integrity is important (and therefore not paying somebody else to do your work for you), but just to remind you:

- it's expensive;
- it's fraudulent;
- you learn nothing about the subject;
- it will be easy for your tutors to spot if there is a significant difference in writing style between the bought coursework and your standard work;
- the penalties are very severe: in many cases this can be automatic expulsion from the university.

Don't do it!

9.5.5 Artificial intelligence

With the rapid evolution of Generative Pre-trained Transformer (GPT) AI resources such as ChatGPT, Google AI, Chatbot GPT it is very easy to get the answers to questions, essays, and reports written with relevant text and including references. This is a revolutionary resource, and it is important that you gain experience in using it effectively as you develop your skill set for your future career. However, there are also dangers. Using GPT solutions to write your coursework will not develop your understanding and ability. You can use GPT as a resource but need to take care to critically evaluate the information it produces as a result of your queries and then use it as you would any other resource: take the information and combine it with your learning from other sources and then write you own answer. That way you will ensure that you build your own understanding of the topic as well as enhancing your skill set.

Remember that simply submitting an assignment written by a GPT system is still a breach of academic integrity.

Some key things to know about Generative AI

1. Hallucinations—it's not always right!

Generative AI uses very convincing language that can feel like an authoritative perspective on a topic, but the facts, points, and content produced are not always correct. This is often obvious to a content expert but not to students of the subject. Always approach outputs with a critical perspective, can you find evidence to support what is being said and is it good evidence?

2. Ethics—we don't know what it has been trained on!

At time of writing, it is unclear exactly what inputs have been used to train the commonly available Generative AIs. This is a problem; the most obvious expression of this issue is the frequent reporting of bias in responses. Uncritical engagement with AI is likely to perpetuate and exacerbate existing social biases. For example, see Zack et al. (2024) on racial bias in medical treatment recommendations (full bibliographic reference at the end of this chapter).

3. Privacy—we don't know where your data goes!

Another aspect of AI that is not clear at time of writing is where your data goes. It is important that you do not share personal information with any AI, we do not know how that will be used in the future or how secure that data will be. How will it be stored, and what might happen if there was a data breach? This could be likened to the use of social media, especially before users were aware of the extent to which data in social media could cause harm or be misused.

4. It has the potential to both enhance and short-circuit your learning!

As we have already discussed getting Generative AI to research and create your assignments for you is academic misconduct and means you will not actually learn much during your degree. If you engage with the outputs uncritically, you run the risk of developing incorrect and uninformed perspectives on the subject it discusses. But if you view the outputs of AI as coming from a talented colleague and use AI to discuss topics and help you sort large amounts of complex information, it can be a helpful starting point. A potentially effective method of practising your critical thinking is to ask the AI about a subject and then see if you can find any evidence to support what it says or see if you can find other perspectives it has not considered.

5. Different rules in each institution

At the time of writing, there has not been a single consistent approach to the rise of Generative AI in Higher Education. Some universities are encouraging staff and students to learn and adapt to the technology, and some universities consider any use of the technology within the assignment process to be cheating. There is also variation in approach between different lecturers on the same course. Best practice is to ensure that students are aware of what they are and are not 'allowed' to do within the rules. Make sure you are aware of the position on use of Generative AI within your university, course, and module before you start the assignment.

6. You should get used to recording your use of AI while creating assignments.

One key element of best practice that is becoming very clear is that people should record and reference how, when, and where AI has been used in the creative process. If you are uncertain whether acknowledgement of the extent of use of Generative AI is going to cause issues with your lecturers, then you should either seek clarification on what is and is not permitted or reduce your usage of the tool in the creation process.

For example . . .

> 'We acknowledge that Generative AIs ChatGPT4 and Google Bard have informed the creation of this section of the chapter through discussions of what students should know about the tool. However, the content is written in our own words and represents a substantial development on the responses to our prompts'

7. It's probably going to be important in your future.

Something else that many people agree on is that Generative AI is likely to become a commonplace tool in your professional and personal futures. It is worth taking some time to familiarise yourself with what it is, what it can do, and how to use it. Used effectively, it can dramatically speed up the creative process but, as with any other tool, its effectiveness is dependent upon how well you can use it. Remember to engage with it critically, carefully, and with integrity!

9.6 Accidental plagiarism and how to avoid it

Hopefully, you now understand what contributes academic misconduct, the things you definitely shouldn't do. However, in our experience, plagiarism is much more likely to happen by accident, through poor academic practice, so let's consider how to avoid that.

9.6.1 Ignorance

Not that you will have this excuse after reading this chapter, but ignorance is the most basic excuse. 'I didn't know' is a common response to accusations of plagiarism, particularly from undergraduates in the early stages of their degree course. However, most universities teach about good academic practice very early in their programmes, so claims of ignorance are not likely to be treated very sympathetically.

How to avoid it

Attend all of your taught sessions and read the university guidance on referencing and academic misconduct.

9.6.2 Poor planning

Poor planning or failing to manage your time well may seem unrelated to plagiarism, but they are, in fact, frequent causes of it. Most students would be shocked at the idea of deliberate plagiarism, especially in a premeditated manner, but when you are running out of time to complete an essay the temptation to take short-cuts becomes stronger. Read the case study that follows and see if you can appreciate what we mean.

Case study—Steve:

> Steve was busy with another assignment and then went out to celebrate when it was finished. Waking up in the morning he remembered that he had a second assignment to hand in and so quickly used Google to search for some information on the web. Using a mixture of copying, pasting, and rephrasing he put together his assignment in a couple of hours. However, his tutors spotted what he had done and he was given a mark of zero

So, plagiarism as a result of poor planning is more plausible than it may at first seem.

How to avoid it

Make sure you have read and implemented the guidance in Chapter 5 *Getting yourself organised*.

9.6.3 Failure to record source details

Failure to record source details is probably the simplest form of plagiarism. It starts at the note-making stage (refer back to Figure 9.2); you find an appropriate piece of text, image, or data that you would like to reproduce or adapt in some way for your essay or report, but you don't write down where the material came from. At the assignment-drafting stage you either forget you got it from somewhere else and believe it to be your own original work, or you re-alise it came from elsewhere but either can't be bothered or don't have the time to retrace the source, so you use it anyway without referencing and hope that no-one will notice. Does this seem implausible? Read the following case study and see if you understand what we mean.
Case study—Andria:

> Andria took a lot of notes while she was researching her assignment, but didn't pay much attention to noting down the referencing details of her sources. When she wrote her assignment she ended up very unsure about what were her own ideas and what she had taken from books. Her lecturer is now talking about plagiarism, but she feels that she was just a bit disorganised and is horrified that she is being accused of dishonesty.

Plagiarism due to failure to record source details is, therefore, a very real problem.

How to avoid it

Implement the guidance in the section above on effective note-making during your re-search process (section 9.4). Pay special attention to the guidance on recording the source.

9.6.4 Inappropriate notes

Inappropriate notes are also a common cause of plagiarism. It is reasonably straightfor-ward to record source details (in theory at least) but how do you know what sort of notes to make on the source? See if the following case study helps you understand what we mean.
Case study—Abdul:

> Abdul has a good understanding of the course, but he finds it difficult to put things into his own words. Academic text books always seem to put things so well that he hasn't got much to add. Sometimes it just seems easier to copy out chunks of different books, paraphrase them a bit, and organise them into an answer. He doesn't feel that this should be considered plagiarism as he has researched the material and put the assignment together from a range of different sources. Unfortunately, his tutors don't agree.

Academic textbooks or journals can indeed 'put things so well' that it's difficult for a student to know what to add. This is why it is so important to work at understanding the concept that you are dealing with, and it takes a certain amount of confidence to do this.

How to avoid it

Implement the guidance in the section above on effective note-making during your research process (section 9.4). Pay particular attention to the guidance on strategic approaches to note-making.

9.6.5 Incomplete referencing

Sometimes students inadvertently commit plagiarism by simply not recording their citations and references fully. This could be due to a lack of understanding of what is required, or a lack of willingness to add what are sometimes just considered to be 'finishing touches', but are in fact crucial to the integrity of their work. Again, see if the following case study helps to illustrate this.

Case study—Jasmine:

> Jasmine worked hard on her essay and was pleased with what she produced. However, she has never understood the point of the referencing system. She included all of the books that she used in a reference list at the end of the essay, but didn't bother citing her sources in the body of the essay. She thought the reference list at the end would be enough, but unfortunately, her lecturers viewed this lack of citing as plagiarism. They covered the assignment in comments and gave her a very low mark.

So, although sorting out the citations and references might just seem like an optional extra, it is vital to get it right.

How to avoid it

Referencing correctly is dealt with in Chapter 11, section 11.2.3.

9.6.6 Lack of engagement with sources

This may possibly be the reason that requires the most thinking to address: it is where a student may well have planned ahead, recorded source details, made appropriate notes, and cited and referenced accurately, but still doesn't demonstrate that they have thought a great deal about what they have written. It is perfectly possible, in an assessed piece of work, simply to string together lots of other authors' ideas (either quoted or paraphrased), but if you fail to include your own ideas, in the form of comment or analysis on the material, you still might be guilty of poor academic practice, even if the source material is cited and referenced correctly.

How to avoid it

Use the guidance in Chapter 10 on how to turn notes into coursework. There is lots of guidance on how to turn the ideas of other people into your own narrative.

9.7 Can you spot plagiarism?

We have now covered in detail the different forms of deliberate and accidental plagiarism and how to avoid them. But can you tell in practice when something is or is not plagiarised?

Try this—Essay extract exercise

Original text

> The Na^+ K^+ ATPase pump has its evolutionary origins in the maintenance of osmoregulation but in the mammalian cell it has become the driver for a wide range of processes, including the generation of the nervous impulse, the absorption of substances by the gut and the re-absorption and excretion of substances by the kidney (Scott et al., 2023).

Essay extract 1

> The Na^+ K^+ ATPase pump has its evolutionary origins in the maintenance of osmoregulation but in the mammalian cell it has become the driver for a wide range of processes, including the generation of the nervous impulse, the absorption of substances by the gut and the re-absorption and excretion of substances by the kidney.

Is this extract plagiarised? Yes or no?

Essay extract 2

> The Na^+ K^+ ATPase pump has its evolutionary origins in the maintenance of osmoregulation but in the mammalian cell it has become the driver for a wide range of processes, including the generation of the nervous impulse, the absorption of substances by the gut and the re-absorption and excretion of substances by the kidney (Scott et al., 2023).

Is this extract plagiarised? Yes or no?

(Continued)

Essay extract 3

'The Na⁺ K⁺ ATPase pump has its evolutionary origins in the maintenance of osmoregulation but in the mammalian cell it has become the driver for a wide range of processes, including the generation of the nervous impulse, the absorption of substances by the gut and the re-absorption and excretion of substances by the kidney.' (Scott et al., 2023).

Is this extract plagiarised? Yes or no?

Essay extract 4

The Na⁺ K⁺ ATPase pump is the driver for a wide range of processes, such as the nervous impulse, the the re-absorption and excretion of substances by the kidney and the absorption of substances by the gut (Scott et al., 2023).

Is this extract plagiarised? Yes or no?

Essay extract 5

The driver for a range of functions such as the initiation of the nervous impulse is the Na⁺ K⁺ ATPase pump, which originally evolved in osmoregulation (Scott et al., 2023).

Is this extract plagiarised? Yes or no?

Essay extract 6

In different physiological systems key functions such as the generation of the nervous impulse and the absorption, re-absorption and secretion of substances are all dependent on the operation of the Na⁺ K⁺ ATPase pump (Scott et al., 2023).

Is this extract plagiarised? Yes or no?

Essay extract feedback

Essay extract 1: This is copied word-for-word but there is no reference. It is a definite case of plagiarism.

Essay extract 2: This is marginally better than extract 1 because the source has been acknowledged by the inclusion of the reference. However, it is still plagiarism: the text is copied word-for-word and should be in quotation marks to indicate that it is the exact wording of the source.

Essay extract 3: Quotation marks are used to acknowledge that words have come from a different source. The quotation marks here make it clear that the student is acknowledging that both the ideas and the words have come from the textbook, so it is not plagiarised. However, stringing together a series of quotations does not demonstrate your understanding of the subject and so is likely to score low marks.

Essay extract 4: This is only a cosmetic alteration. The wording and sentence construction bears very close resemblance to the source and so is plagiarised.

(Continued)

Essay extract 5: This student has just swapped a few words, perhaps using a thesaurus to find replacements, but this is not sufficient to be considered new work. The thinking has gone into finding alternative words to avoid directly quoting rather than into understanding the statement, and so is plagiarised.

Essay extract 6: The student has understood the source and put the relevant information into his or her own words. This demonstrates the student's engagement with the text and ability to explain the information relevant to the essay, and so is not plagiarised.

 ## Chapter summary

In this chapter, we have looked at the approaches to undertaking the research for your assignment and the different types of source material available to you. We have reflected on the importance of making accurate notes and recording the details of the sources. In the final section we considered the significance of academic integrity and good academic practice in your development as a scholar and professional bioscientist, leading on to consideration of what plagiarism is and how to avoid it.

 ## References

International Center for Academic Integrity (2021). *The Fundamental Values of Academic Integrity* (3rd Ed). ISBN 978-0-9914906-7-7.

Zack, T., Lehman, E., Suzgun, M., Rodriguez, J. A., Celi, L. A., Gichoya, J., Jurafsky, D., Szolovits, P., Bates, D. W., Abdulnour, R. E. E., and Butte, A. J. (2024). *Assessing the potential of GPT-4 to perpetuate racial and gender biases in health care: a model evaluation study.* The Lancet Digital Health, 6(1), pp.e12-e22.

Developing visual and oral presentation skills

Introduction

Increasingly, Bioscience students are expected to be able to effectively communicate information to a variety of audiences using a diverse portfolio of digitally enhanced formats. Coursework assessment of Bioscience degrees commonly includes written essays and reports (as covered in Chapter 8), as well as posters, infographics, and verbal presentations. Some degrees even include podcasts, videos, policy briefs, debates, and many other modes of communication. We define these different types of assessment as their 'format'. The format is concerned with how you are expected to communicate your key points and what audience you are expected to communicate them to. During your course, you will probably be expected to achieve similar assignment aims across multiple different formats. For example, from a single laboratory practical, you could be asked to present your results as a scientific report (written), poster (visual), or presentation (oral). Sometimes, the course may be designed to use one format as an opportunity for feedforward for a different format using the same set of information. For example, it is very common for final year students to do a poster for their final year projects/dissertations that feeds into the final written report. Given the variety of assignment formats you could be set during your course, we have tried to provide guidance and techniques that can be applied to many different coursework formats (as opposed to covering each in turn and repeating a lot of the guidance). This chapter focuses on developing the visual and oral communication skills you will need to create coursework in formats other than long-form written assignments.

10.1 Before you start

In the previous chapters, you have analysed the assignment briefing provided to you and hopefully identified one of two assignment aims: 1) independent investigation and discussion of a topic (e.g. an essay), or 2) collect, present, analyse, and discuss data in some form of practical report. If you have followed the process outlined in the book, at this stage, you will have a plan that outlines the key topics that you want to include in your assignment, notes of the independent research that supports your key points, and the order that all this needs to be communicated in. This chapter focuses on how to create a variety of different coursework formats, but it largely assumes that you have a solid plan of what content needs communicating and what order to communicate it in. If you have not already analysed the assignment brief, researched the topic, and created a plan, it might be a good idea to go back to the relevant chapters and complete those tasks first. However, if you just want to know what a good poster looks like, or get tips on confident public speaking, (or a similar kind of query) then you can jump straight to the relevant section in this chapter.

10.2 Creating assignments focused on visual communication (e.g. posters and infographics)

Most of your lecturers will have created a 'research poster' during their career. This most commonly happens at a conference, which is an event where many researchers congregate to present and discuss their work. Conferences are important places to:

- publicise the findings of ongoing experiments to the research community;
- discuss past, current, and future research projects with other subject experts;
- create opportunities for collaboration with other researchers from other institutions and countries (or even other disciplines!).

Often there is not enough time for everyone at a conference to be able to present their research orally, so poster presentations are commonly used to allow lots of people to present their research at once. Typically, posters are size A1 (eight times bigger than A4) or A0 (16 times bigger than A4) and are attached to display boards arranged throughout an open plan room or conference hall. Time will be allocated in the conference schedule for delegates to view the posters, usually with opportunities to discuss the work with the presenter. A poster presentation is more limited than an oral presentation in terms of the amount of information that can be communicated and, therefore, posters should be designed to summarise the important elements and promote discussion. This is a difficult and important skill that can be transferred to many contexts after you graduate.

Because of this experience above, the 'research poster' has been a staple of assessment in Bioscience degrees for some time. It will typically require creating something like in Figure 10.1.

The poster in Figure 10.1 has been designed to summarise and visualise the key findings of a specific research study. Note that it includes the sections we would expect to find in a scientific report (see Chapter 8, section 8.3). This format will probably work well for assignments where the aim is to 'collect your own data, present and analyse the results, and interpret them in the context of the research literature' to a specific scientific audience (as is the point at a research conference). However, increasingly, Bioscience degrees require students to create poster-like coursework for a variety of aims and audiences. For example, you might need to blend visual design and written essay skills to create a policy brief document designed to inform a member of parliament on the science underpinning potential solutions to a societal challenge (e.g. climate change, healthy diets, or air pollution). This chapter will attempt to cover some basic principles that can be applied to different coursework assignments that require visual design skills to help communicate clearly, concisely, and engagingly with different audiences.

10.2.1 Which software?

One thing that may become quickly obvious when starting an assignment that requires visual design skills is that you have many more software options compared to text-based assessment, which will mainly use one of the Microsoft, macOS, or Google word processing

Figure 10.1 Research poster template in a format that one might expect to find at a research conference. Note, the template includes some guidance as to what one might put in each section of the poster, and you could use it to help structure your own research poster.

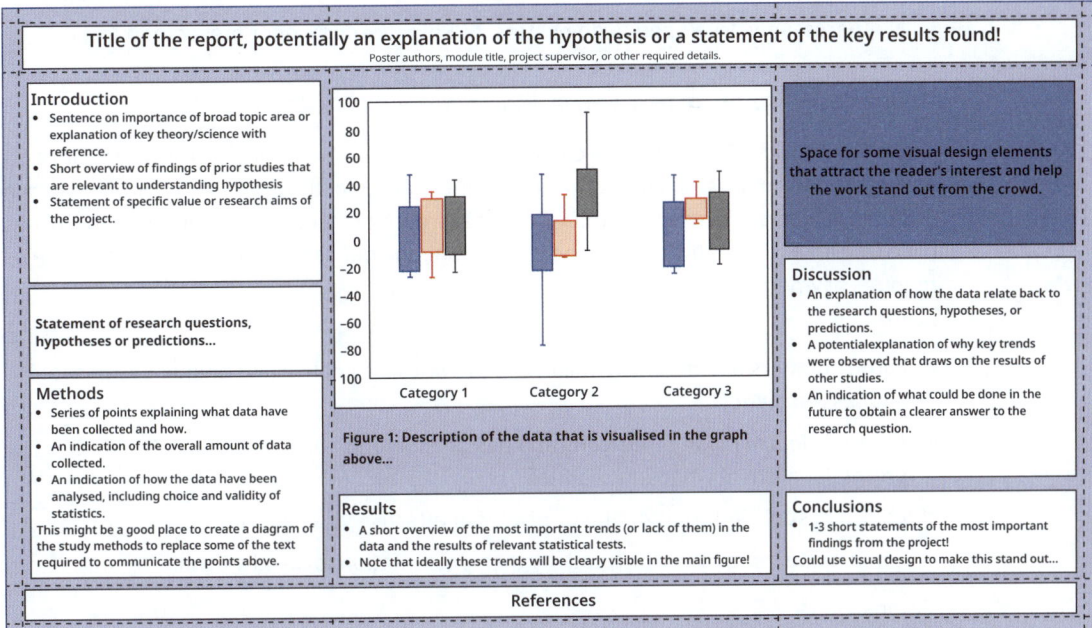

software (which are largely identical). When it comes to visual design elements, you may already be familiar with the presentation software from these big three companies (e.g. Microsoft PowerPoint), but there are also some specialist desktop and online software options that can encourage more creative designs. This section is not designed to be a detailed investigation and tutorial in the use of these different options, but we would like to make you aware of what is available and some pros/cons as we see them.

Microsoft PowerPoint, Google Slides, and macOS Keynote

These options are mainly focused on creating slides to support oral presentations, but they can be repurposed to create posters and infographics.

Pros:

- Students tend to be familiar with the basics of their use because of the unified graphical user interface across the different operating systems (i.e. if you know how to use Word, you can probably navigate PowerPoint).
- They are simple and there are lots of online tutorials on how to push the boundaries with their design capabilities.
- Templates available that provide examples of different designs and layouts.
- Often easy to collaborate using cloud storage. Commonly available on university-managed computers.

Cons:

- Require some initial set-up in order to create posters/infographics instead of presentations.
- Often limit design creativity a bit and can be difficult to create a professional finish.
- Not necessarily developed to do posters, but highly effective for presentations.

Online design software (e.g. Canva, Biorender, and Piktochart)

These are websites that have been created to help people who do not have a particular background in design develop creative, well-designed, and mostly professional products (such as infographics, social media banners, and classroom posters). As of writing, the websites have free access accounts that can be used by anyone but access the full suite of resources requires payment.

Pros:

- Large numbers of templates available to provide imaginative and unique designs for your assessment.
- Access to lots of royalty-free icons and design elements that can enhance professionalism of what you produce.
- Lots of bespoke tutorials on the principles of effective visual design.
- Can provide more room for creativity than presentation focused software.

Cons:

- May need to pay for access to full suite of resources (although the free version should suffice).
- Some learning curve to develop familiarity with how they work.
- Can be challenging to develop a final polished product with a fully professional finish until more familiar with techniques.

Adobe software (e.g. Illustrator or InDesign)

These are specialist software commonly used by visual designers to produce highly professional products for a wide range of aims and audiences (please note other similar software are available).

Pros:

- Flexible and can be used to produce products with a highly professional finish (e.g. ensuring that picture backgrounds have been removed to blend more seamlessly into the overall design).
- Highly in-demand transferable skill that would set you apart from other Bioscience graduates.

Cons:

- Relatively large learning curve before they can be used effectively.
- May not be available through university-managed computers and individual licenses are comparatively expensive.

Whichever software you decide to use, it is useful to make sure you use the tutorials available online (and potentially from your university library) to make the most of it. Your choice is likely to depend on how much time and energy you want to invest in developing your visual design skills. You may see this as an opportunity to develop an important skill set that will really add value for future employment (e.g. designing social media campaigns as a conservationist or health professional), in which case this is a great time to become familiar with some of the more advanced software.

10.2.2 The overall design—professional layouts

Once you have picked your software, it's time to actually create the assignment. The overall layout of your product is often a great place to start, especially if you already have an assignment plan with the key sections/topics that you need to cover (as per Chapter 8). An effective layout is one that has been designed to help the person viewing your product to understand the sequence of the information (visual hierarchy). 'Visual hierarchy' is the term design professionals use to describe the order in which the reader needs to navigate the visual content. A combination of a clear layout and appropriately formatted headings, as shown in Figure 10.1, is the best way to establish a visual hierarchy.

It can be helpful to start your design process with a blank design grid, as shown in Figure 10.2. A design grid is simply a series of horizontal and vertical lines on the page that are used for lining sections up; the lines are then removed when the poster is printed. Most drawing software has a design grid feature, or alternatively you can draw the lines manually with a line tool to create the grid and remove them before printing. Alternatively, you can use a template to provide the overall layout structure (there are lots of templates available when using software such as Canva). Generally, it is good practice to have an overall layout that has a number of boxes that you can put text in (as opposed to placing text directly on the background) and space for some sections with visual elements (e.g. graphs). It is important at this stage that the number of sections within your layout aligns well with the number of topics you need to cover in the assignment.

Aligning layouts to audiences

Scientific research posters tend to have a strongly aligned and blocky layout with a lot of structure. The sections should match closely with the sections of a scientific report (so make sure to check back with Chapter 8, section 8.3, to ensure you understand what kinds of information need to go in each section). In order to make the most of the poster format, you should ensure that the visualisation of your data is appropriately large, easily legible, and really helps the reader understand exactly what your study has found. You should use bullet points to indicate what you might say only, and you can adapt them to fit your needs. This approach is illustrated in Figure 10.1, which includes some guidance on the kind of content that might be expected in a research poster.

Compared to a scientific report poster, an infographic or visual format focused on a broader audience can work best with a slightly less blocky layout. Note the fundamental grid would be the same for this product as the scientific poster style one, but elements don't have to align to it so rigidly. These kinds of product generally have more room for student creativity, and benefit from consistent attempts to convey information through text and visual design simultaneously. Taking the time to create your own diagrams can really help

Figure 10.2 Some example layout grids that create different numbers of spaces and relative sizes

evidence detailed understanding of the subject in very few words. You can use the SmartArt tool in Microsoft PowerPoint, which has a large number of templates to work from. However, with a slightly relaxed layout you need to make sure you are consistent with your use of visual design to guide the reader through the information.

It can be easy to make your layout too busy and too unstructured, which means the reader will lose interest in your product quickly. For example, a very minor change in layout between two figures might result in a product that is almost impossible to follow, uses space too inefficiently, and looks less professional. Remember most of your audience will instinctively read the content from top to bottom and left to right.

Key points on picking a product layout:

- Overly structured products can fail to catch the attention of the audience (especially if there is a room full of identical posters layouts with similar colour palettes).
- Products with insufficient structure look unprofessional at best, and at worst can make it impossible for the reader to quickly and efficient extract key information, which means that the person grading your work may not be able to find the content they are looking for even if it is there.
- The appropriate level of structure is commonly related to the audience for the product:
  - More specialist audiences need to grapple with more complex content and may benefit from more defined structures (e.g. a research poster).
  - More generalist audiences tend to place more emphasis on visual interest and engagement with a small number of key messages (e.g. a behaviour change infographic).

Try this—Designing the layout

1. Try a couple of the different design grids and templates from those available. Make sure there are enough textbox spaces for each of the sections in your plan (e.g. Introduction, Methods, Results, Discussion) AND space for some relevant visual elements!
2. Add the subheadings to each of the text box sections.
3. Add a sentence for each key topic that needs to be included under each subheading.

Things to be aware of:

- Is the order in which the reader should approach the information obvious and intuitive?
- Is there a good balance of content space and neutral space, so that the content does not look crowded.
- Is there plenty of space for non-text elements (i.e. figures, graphs, diagrams, etc.)?

10.2.3 Formatting the type

Before you start adding the final content, it is a good idea to format the type. This will give you a clearer idea of how much space you have to work with, as you will know how big your text will be. When formatting type for a poster, you need to remember two important principles:

1. Use consistent styles;
2. Group sections of text appropriately.

Use consistent styles

Using consistent styles helps achieve what graphic designers call the principle of *resemblance*. It means that text which performs the same function should always be formatted in exactly the same way. For instance, if you decide that the subheadings should be Arial, 10 point, bold, left-hand justified with an after-paragraph line spacing of 0.2 lines, then all your subheadings should be formatted in exactly this way. Similarly, you might decide that the captions for figures should be Arial, 6 point, regular, right-hand justified with a before and after paragraph line spacing of 0.2 lines, in which case all your captions should be formatted in exactly this way (see Table 10.1).

There are two reasons for using consistent styles: first, it makes the poster look neater; secondly, it makes the structure of the poster easier to understand, because you know which text is performing what function. An example of using consistent styles is shown in Figure 10.3.

Table 10.1 Suggested font sizes to allow appropriate magnification

Style	A4 to A1	A4 to A0
Title	18 point Arial bold	16 point Arial bold
Heading	13 point Arial bold	12 point Arial bold
Subheading	11 point Arial bold	10 point Arial bold
Introductory text	11 point Arial bold	10 point Arial regular
Body text	10 point Arial bold	8 point Arial regular
Captions	8 point Arial bold	6 point Arial regular

Figure 10.3 Using a limited number of consistent styles

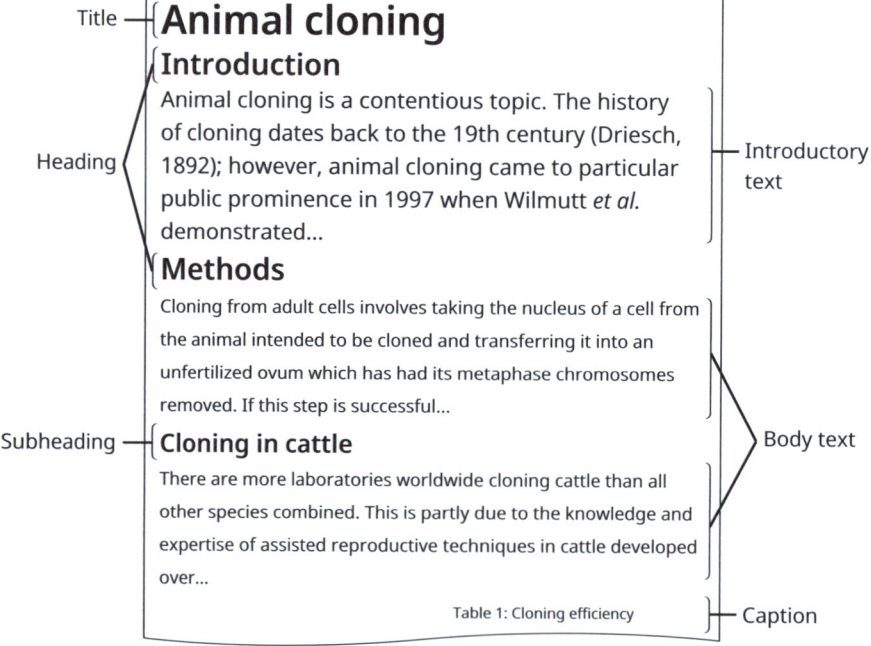

Group sections of text appropriately

Grouping sections of text appropriately helps achieve what graphic designers call the *principle of proximity*. It means that text which belongs together is grouped together; this is usually achieved using paragraph spacing. Figure 10.4 illustrates inappropriate spacing of text; there is a whole line space between each section and in the bulleted list. The result is that the headings and bullets float around and it is not immediately clear which piece of text belongs to which section.

Figure 10.5 illustrates more appropriate spacing of text; paragraph spacing has been used to make it clear which pieces of text the headings and bullets belong to.

Figure 10.4 Inappropriate proximity of text

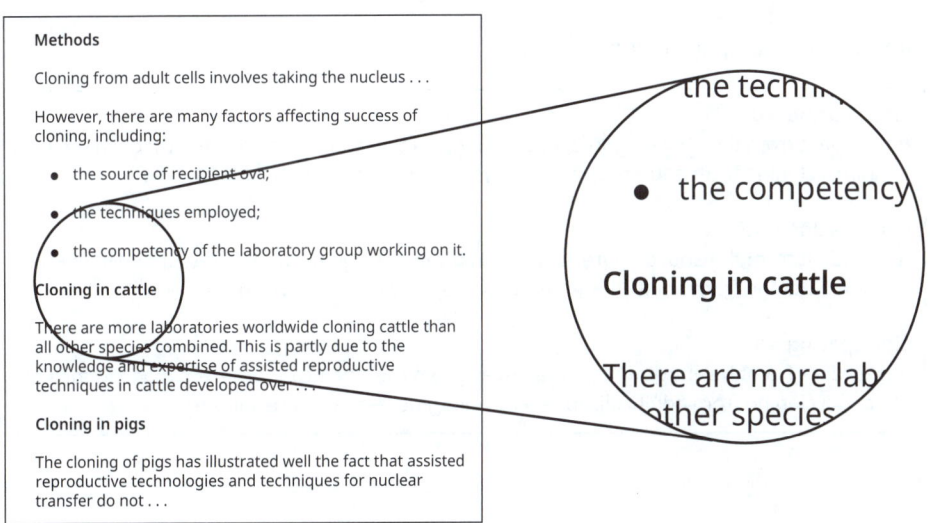

Figure 10.5 Appropriate proximity of text

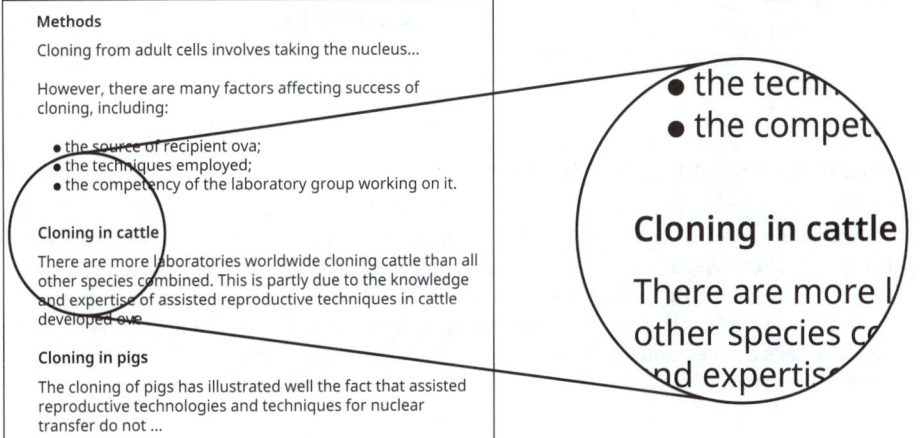

Line spacing, alignment, and text formatting

In addition to thinking about the principles of *resemblance* and *proximity* when formatting the type, it is also important to consider line spacing, alignment, and text formatting. Figure 10.6 illustrates the effect of line spacing on body text; body text (which is usually the smallest text on the page other than captions) is much easier to read if the line spacing is slightly increased.

Titles that run to more than one line, however, often benefit from the reverse: compressed line spacing, as shown in Figure 10.7. However, where possible, it is better if you can just make your titles more succinct so that they can fit on one line.

It is recommended that you left-hand justify all body text (as shown in Figure 10.8) because fully justifying the body text can result in awkward gaps on some lines, making the text more difficult to read (as shown in Figure 10.9).

Figure 10.6 Line spacing and body text

Line spacing = 0.9
It is evident from the above figures that, even given the significant progress made in the field of animal cloning from adult cells in recent years, success rates are still very low.

Line spacing = 1.0
It is evident from the above figures that, even given the significant progress made in the field of animal cloning from adult cells in recent years, success rates are still very low.

Line spacing = 1.1
It is evident from the above figures that, even given the significant progress made in the field of animal cloning from adult cells in recent years, success rates are still very low.

Figure 10.7 Line spacing and titles

Large text (e.g. titles) probably needs decreased line spacing . . .

Line spacing = 0.9

. . . because it looks a bit strange with increased line spacing.

Line spacing = 1.1

Figure 10.8 Left-hand-justified body text

Animal cloning is a contentious topic. The history of cloning dates back to the 19th century (Driesch, 1892); however, animal cloning came to particular public prominence in 1997 when Wilmut *et al.* demonstrated that the cloning of mammals from adult cells, rather than an embryo, was possible. The authors of the study derived nuclei of cultured mammary epithelial cells from an adult ewe, resulting in the birth of 'Dolly'—the first cloned sheep.

Figure 10.9 Fully justified body text

Animal cloning is a contentious topic. The history of cloning dates back to the 19th century (Driesch, 1892); however, animal cloning came to particular public prominence in 1997 when Wilmut *et al.* demonstrated that the cloning of mammals from adult cells, rather than an embryo, was possible. The authors of the study derived nuclei of cultured mammary epithelial cells from an adult ewe, resulting in the birth of 'Dolly'—the first cloned sheep.

Finally on formatting the type:

- use the same font throughout your poster (unless using a complementary font for headings);
- set the headings in bold;
- use italics, underlining, and CAPITALS sparingly (only when convention dictates);
- break up any large areas of text with subheadings.

Try this—Formatting the type

If you completed the previous 'Try this' activity, you will have a draft poster with subtitles and key topic sentences in each textbox space. Next you should:

1. Pick an appropriate font style and size (formatting) for your title, subheadings, and main text.
2. Format all of the draft text on your poster to one of these three formats.

Things to be aware of:

- Is the order in which the reader should approach the information obvious and intuitive?
- Is there a good balance of content space and neutral space, so that the content does not look crowded?
- Is the text legible from at least a couple of metres away in its final print size?
- Is it clear which pieces of text belong to which section of the poster (text boxes are often the best way to achieve this)?

10.2.4 How much content?

Now we have the overall design outlined, you need to add the actual content. A common part of the marking criteria for assignments that require visual design is related to the amount of text on the final product. Too much text defeats the purpose of the assessment format, by overwhelming the reader and making it harder to extract the key points. Alternatively, too little text commonly means that the subject is not covered in sufficient detail. Writing concisely and strategically is a difficult skill and is a core component of doing well in these types of assignment. It is also important to note that effective visual design requires you to communicate information visually (and not just through text), which can really help with minimising text on your final product.

Figure 10.10 A poster with 300 words

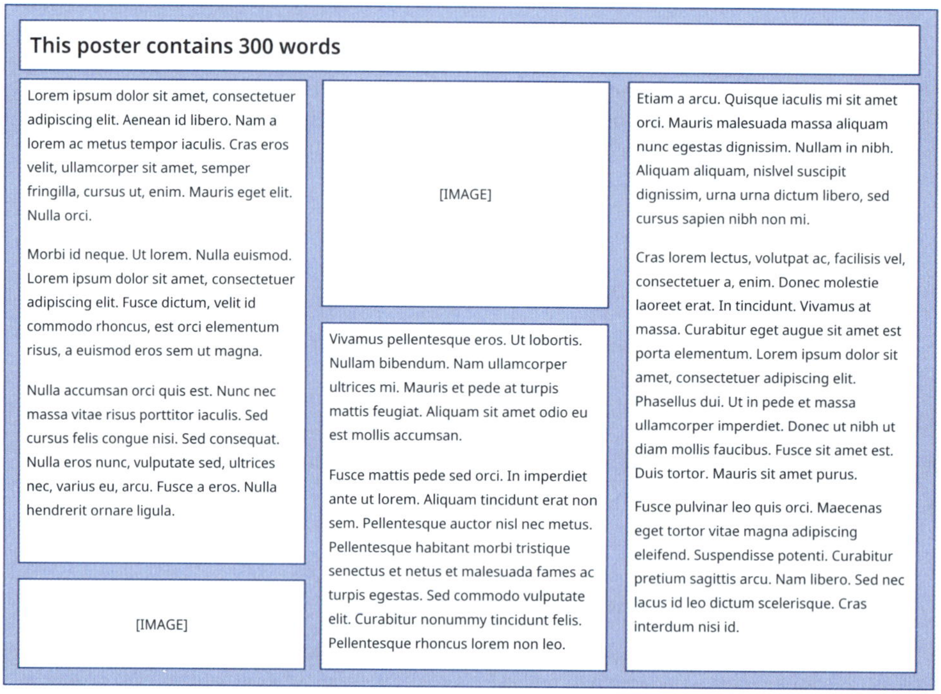

Traditionally, the recommended number of words for a research poster is between 300 and 400, but modern design trends are pushing this number lower. Someone who is really good at visual design could produce high-quality research poster with fewer than 250 words. An infographic or other product aimed at a more generalist audience could push this number even lower as reducing the number of words can really help emphasise a smaller number of key messages. We have outlined what each of these might look like in Figures 10.10 and 10.11.

Ultimately, your final product will need to balance detailed coverage of the subject without cluttering your final design with too much content. In order to achieve this, you will need to write concisely and strategically:

Writing strategically

As a general rule for assessment, the shorter the word count, the more essential it is that you know what content you need to include in your assignment. While this is important in every assessment format, it is crucial to performing well in visual design-based assessments. Having a good plan can really help with this because it means you know exactly what content needs to be in each section. Make sure you have analysed the assignment brief (including the marking criteria, as per the guidance in Chapter 7), know what the aim of the assignment is, and have read the guidance on what to include (e.g. what should go in

Figure 10.11 A figure with 500 words. Please note that for most modern science posters this would be too much text, if you find yourself with a poster that looks like this it would be a good idea to think about how you can make the content more concise or to communicate some of your concepts through diagrams and illustrations

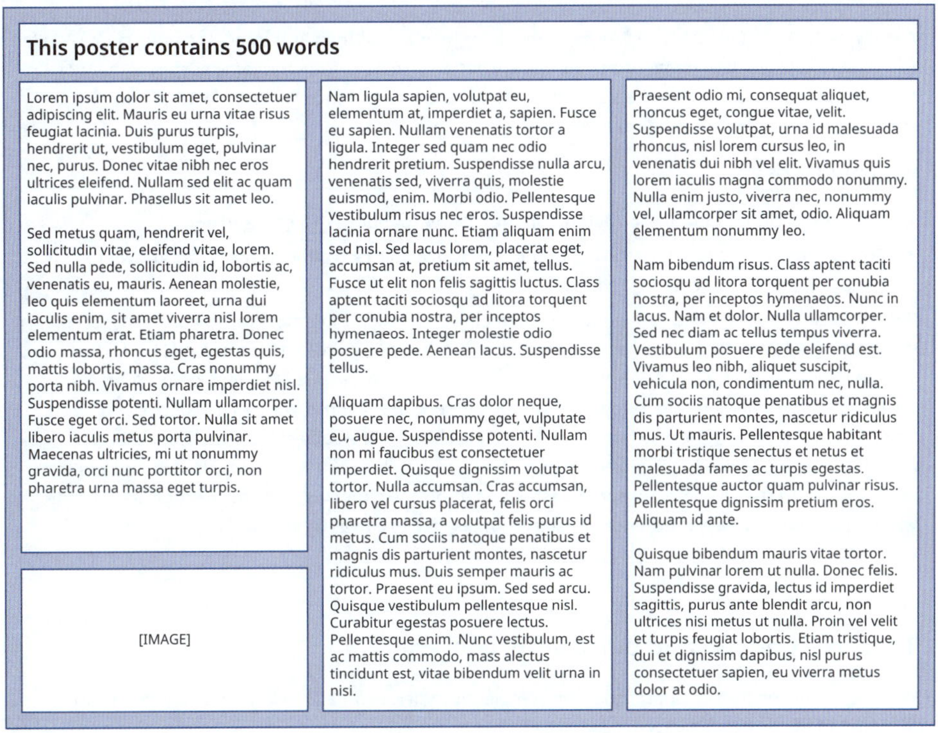

each section of a scientific report: Chapter 8, section 8.3). You can then have a short bullet-pointed sentence for each of these strategic needs, which will help ensure that no text is superfluous and the reader (by which we mean the grader) can find what they are looking for easily. This is most likely to closely resemble the plan you will have made if you have been doing each chapter in order.

Writing concisely

- Bullet points are a common method of concise writing.
- They dispense with the need for prose, which reduces word count.
- They help provide order, structure, and spacing, which improves clarity.
- Consider using bullet points in your final products.

Concise writing does not mean that you need to reduce engagement with appropriate literature. High level scientific writing requires you to synthesise key points from across multiple sources.

Either way, it is important that you try to think about exactly what it is that you want and need to say, and then try to refine that into as few words as possible!

> **BOX 10.1 An example of how the same content might be written for an essay compared to a poster**
>
> **Instead of ...**
>
> A study by Rex et al. (2012) found that dogs are better than cats as long as they are the same colour. While Felix and Felix (2020) found that cats in Europe are better than dogs from the same continent, but not from different continents. Conversely, Rattus (2021) found no difference between cats and dogs in their study focusing only on small cats and dogs.
>
> **We could write ...**
>
> - Prior research has found inconsistency in the relative 'goodness' of cats and dogs.
> - Colour (Rex et al., 2012), location (Felix and Felix, 2020), and size (Rattus, 2021) are known to be important determinants of 'goodness'.
>
> Both of these texts show relatively equal engagement with available literature (an important aspect of determine grades in many instances) but the first version uses 65 words and the second 35. The first version might be an appropriate level of detail for an essay where we need to know some additional details of exactly what the studies found. However, the second version is much more concise and would be sufficient in a poster where we are trying to establish that (a) the subject has been studied before, (b) that there is no scientific consensus on the subject, and (c) there are some variables that are known to be important. **In the case of a research poster, you might be expected to know the details of studies you cite if there is a poster defence/presentation aspect to the assignment.**

Try this—Concise and purposeful content

Working within the textbox spaces that you completed earlier, rewrite your key topic sentences into a first draft of the content. Things to be aware of:

- Ensure that the text covers the key points required to achieve the aims of the assignment in concise, easy-to-understand sentences.
- Make sure that the text remains in the format that you originally set it (remember three distinct formats: title, subheading, and main text).
- Is the text legible from at least a couple of metres away in its final print size?
- Is it clear which pieces of text belong to which section of the poster?
- Are there large blocks of unbroken text that make the poster look crowded, or make your key points difficult to find?
- Do the text fit neatly within the textbox spaces allocated to them?

10.2.5 Using visuals effectively

It is important to remember that posters, infographics, and the like are fundamentally visual communication formats. Text is necessary and essential, but many students make the mistake of relying on it too heavily. Effective use of visual elements can:

- help keep word counts minimal, which in turn emphasises key messages in the text;
- add interest and engagement to the overall design;
- communicate complex ideas more effectively by engaging multiple parts of the reader's brain.

In short, visual elements are a fundamental reason for using this format of communication. So how can you use them effectively? Well different types of visual element can be used in different ways.

Diagrams

A well-chosen diagram can save a lot of words or make text descriptions much easier to understand. Diagrams can be an excellent way to achieve all of the benefits that visual elements can bring. In fact, we strongly encourage you to try making your own diagrams if you cannot find one that does the exact job you want it to. Designing your own diagram often means you have more capacity to align it to the design decisions you are made in your poster/infographic. Some ideas about places where diagrams are commonly particularly helpful:

- Visualisations of your study methods, such as the sampling strategy, use of key apparatus, or data categorisation process (e.g. an ethogram of behaviours observed).
- Explanations of key theories or scientific processes that are important for understanding the poster contents (e.g. they might underpin your study hypothesis and predictions).
- Frameworks that explain how multiple different concepts fit together, or how your results relate to real-world processes.

As suggested in Figure 10.1, methods sections are often wordy and difficult to explain to other people, which makes them a great place for a simple diagram.

Photographs

There are three key points you need to remember about any photographs you use on a poster/infographic.

1. They need to be relevant to the content of your poster, not just put in to brighten it up.
2. If they are not your own photographs they need to be appropriately referenced (in the same way as you would reference text that wasn't your own).
3. They need to be of an appropriate resolution.

In order for photographs in particular, and images generally, not to degenerate, they need to be of sufficient quality to start with—that is, they need to be of a sufficiently high resolution. The resolution of digital images is measured in pixels per square inch. Photographs copied from the internet usually have a resolution of 72 pixels per square inch and so will look 'grainy' or 'pixelated' when enlarged, as shown in Figure 10.12.

For a printed poster, choose images that have a resolution of at least 300 pixels per square inch at the size at which they will appear on the poster. In all likelihood, you will want to enlarge the image when you reproduce it for the poster: and as soon as you enlarge it, the resolution will decrease. For example, if you have an image that is 2 cm square and has a resolution of 300 dpi, when you enlarge it to become 4 cm square, its resolution will drop

Figure 10.12 The effect of enlargement on low resolution images

to 150 dpi (which is too low a resolution for printing purposes). Instead, start with an image that has a resolution that's well above 300 dpi. Then, when you enlarge it, the resolution will drop—but hopefully not below 300 dpi.

A common issue found in many posters is the use of pictures that are not effectively blended into the overall design. Examples of this might be a white, square background box on top of a background colour. Another example is the use of a very busy, and low resolution image as the product background. Unless you are very confident in what you are doing, plainer backgrounds are easier to work with.

Graphs

When making graphs for posters, you need to think carefully about how you format them, as viewing them from a distance means that default formatting is usually not appropriate. The graph in Figure 10.13 is difficult to view from a distance because of the relatively small font, the shaded plot area and the excessive use of horizontal lines. All these formatting features are default features in Microsoft Excel. As a general rule, when it comes to the formatting of a graph, don't accept the default formatting provided by your software.

Figure 10.14 shows a more appropriately formatted graph. The font size has been increased, the shading on the plot area has been removed, and the use of horizontal lines has been minimised. Additionally, the colours of the bars have been altered to provide greater contrast, making the two sets of figures easier to distinguish between. Again, as with photographs and diagrams, if the data are not your own, you need to reference them appropriately.

Figure 10.13 An inappropriately formatted graph

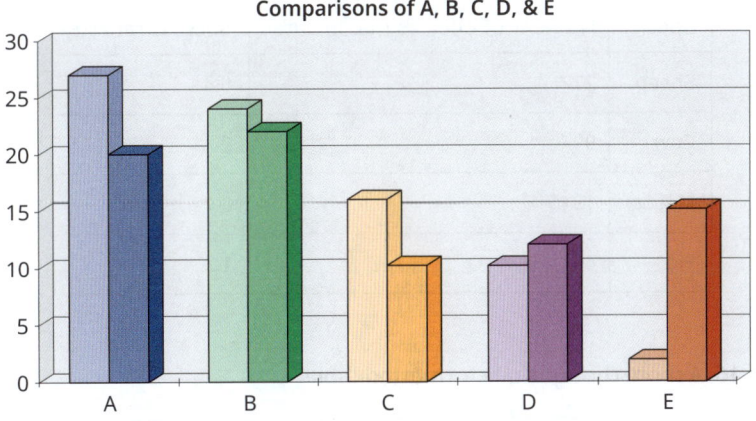

Figure 10.14 An appropriately formatted graph

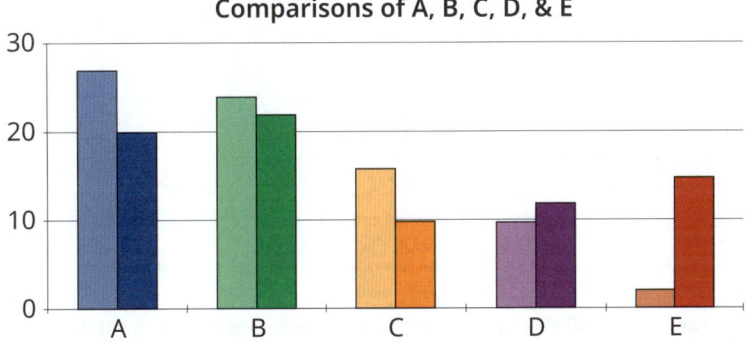

Tables

Finally, in this section on preparing images, let's consider tables. The principles that applied to graphs apply equally to tables—format them in such a way so as to make them clear, simple, and easy to see. Again, default formatting is usually inappropriate, as it tends to make the text too small and the gridlines can make patterns in the data difficult to discern. As a general rule, don't format tables that look like spreadsheets, as shown in Figure 10.15, as they are more difficult to interpret, even when you are dealing with a small data set.

Figure 10.16 shows a table that has been formatted more appropriately. The following changes have been made:

- the font size has been increased;
- the header row is made clear by setting the text in bold and lightly shading the row;
- the internal vertical lines have been deleted to allow patterns between the rows to be more easily seen;
- the numbers have all been right-hand justified so that the decimal place is vertically aligned;

Figure 10.15 Default formatted table

Species	Number of oocytes manipulated	Success rate (%)
Sheep	277	0.4
Cow	99	2
Mouse	1345	0.7
Pig	401	1.2

Figure 10.16 A more thoughtfully formatted table

Species	Number of oocytes manipulated	Success rate (%)
Cow	99	2.0
Pig	401	1.2
Mouse	1345	0.7
Sheep	277	0.4

- the percentages have all been quoted to the same number of decimal places;
- the rows have been ordered by the magnitude of the aspect of the data that attention needs to be drawn to: the percentage success rate.

How to use visual elements

The exact visual elements you include will depend on the aims and content of the product you are making. Next, we will have a look at some general rules and patterns you can use to include them in your product.

Assignments aligned to scientific reports—these should have at least one focal visualisation of your data or key finding. A common error with research posters is that they make their key figure too small in order to include more text. The graph or figure that visualises your data should be a key focus of the overall design. As outlined above, it can also be helpful to think about creating a diagram to help explain your methods.

Assignments discussing scientific concepts—formatting these according to a diagram that explains the concept is often helpful. Whenever you are trying to explain a complex concept, making your own diagram can help you communicate ideas concisely and effectively.

Assignments making an argument—arguments are best made with data and evidence. Including some figures or table that visualise relevant data and linking them to the key points you need to evidence allows you to reinforce key points as well as add visual interest to the product.

Try this—Communicate visually

Reflecting on your design so far:

1. Add some key visual elements that will help you achieve the assignment aims (e.g. figures of your data).
2. Identify spaces in your design that have a lot of text. Consider if some of that text would be more concise if you had a visual element to support communication of the key points.
3. Are there any large areas of blank space that would benefit from some visual elements to add engagement?

Things to be aware of:

- Is there a good balance of text, neutral space, and visual design elements?
- Are the visual elements of sufficiently high quality (e.g. high enough resolution)?
- Are the visual elements sufficiently large such that all of the details can be seen?
- Do the visual elements fit well into the overall design?
- Do all of your visual elements clearly link to the text (e.g. is there a sentence summarising the key finding from each figure?).

10.2.6 Colour schemes and encouraging interest

The colour scheme is a core aspect of the visual design of your product. It will do a lot of heavy lifting when it comes to engaging the reader, highlighting key messages, and standing out from a crowd of similar submissions. However, it can be easy to overdo this. Try to use only two or three different colours, plus black and/or white. There are two ways of choosing your colours: you can use colours from your images or a colour wheel.

Colours from visual elements

One way of selecting colours is to use one that is represented in your photos, diagrams, and graphs. So, for the poster we have used as an example we might choose a pink because of the pig photograph, or a brown because of the cow photograph.

Colour wheel

The alternative to choosing colours from images (either because you don't have any appropriately coloured images or you are not confident in choosing from them) is to choose colours from a colour wheel. A colour wheel shows the primary, secondary, and tertiary colours, and can be used to choose appropriate mixes of colour.

You can choose either:

- analogous colours (colours next to each other on the colour wheel);
- complementary colours (colours opposite each other on the colour wheel);
- or shades (different shades of the same colour)

Try this—Designing the colour scheme

If you have followed the steps so far, you should have most of the content, layout, and formatting for your product. Now try to apply a colour scheme of your choice to the content. If there are dominant colours in any of the visual elements, then you could try aligning the scheme to those. Otherwise, select some from the colour wheel ideas above.

Things to be aware of:

- Is there sufficient contrast between your text and the background it is sitting on, so that it is easy to read at a distance?
- Have you got a sensible number of colours that work together (most likely no more than three)?
- Is your product accessible (e.g. green/red contrasts are generally considered inappropriate)?
- Is your product likely to stand out among a row of similarly designed products?

10.2.7 Check it very carefully

Chapter 11, section 11.1.1, on the review and redrafting process, covers this in more detail but there are some specific checks that you can perform for visual design-based products (Box 10.2). This is particularly important if you are going to get an enlarged, laminated copy printed: you don't want to spend money printing it out only to find the errors afterwards. A good way to check your poster is to print it out A4 (preferably in colour), stick it on the wall, and take a step back to look at it.

This will simulate more closely how the real thing will be viewed and help you find any errors more easily. Alternatively, if you have access to a data projector, you can project the

---

BOX 10.2 Do the small details really matter?

Posters can be very fiddly and therefore time-consuming to create. However, remember the academic context of poster presentations that we summarised in section 10.2 *Creating assignments focused on visual communication*: in a busy conference setting with lots of posters to view, people make very quick (often unfair) judgements about whether they will bother viewing a poster. We said that a poster needs to stand out from the crowd in some way if it is going to be viewed. Positively, this will include a snappy title, good visual impact, and information that is readily understandable. Negatively, small errors in the design can put people off and lead them to think that, if the design isn't accurate, the data might not be either. Also, if you are creating a poster at A4 size, then enlarging it to A1 or A0, any small errors at A4 will be enlarged 300% or 400%, respectively. Therefore, attention to detail is important. See Figure 10.17 for an example of how small inaccuracies with the layout at the A4 stage lead to much larger inaccuracies at A0.

(Continued)

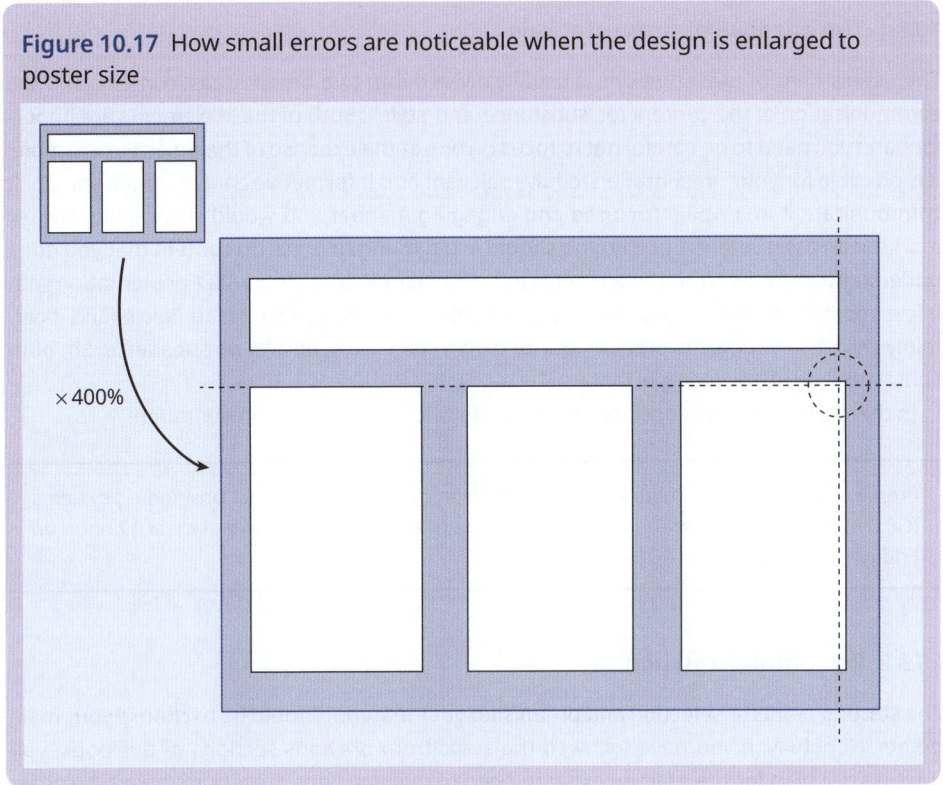

Figure 10.17 How small errors are noticeable when the design is enlarged to poster size

× 400%

image onto a screen to see the effect of the image being blown up to full size (although the image resolution is much less for digital images than printed images, so don't use a projected image as a guide for image quality).

10.3 Designing and delivering scientific presentations

The ability to present information orally to a group of people is an increasingly important skill for students undertaking Bioscience degrees, or any other kind of degree for that matter. It's also one of the key skills that many employers look for across a wide range of roles. Some students find the prospect of standing up in front of a group of their fellow students and academic staff to give a presentation is the stuff of nightmares, even more so when the presentation is assessed. For other students, presentations are not quite such a frightening prospect, but even so, the public nature of a presentation, compared with the privacy and anonymity of an essay or exam, for example, is a factor that makes getting presentations right all the more important; no one likes being embarrassed in front of their peers.

This section starts with some guidance on creating presentation aids (e.g. PowerPoint slides) to support a traditional scientific presentation assignment. However, the focus of the section is on preparing and delivering confident oral presentations, which can be applied to a variety of assessments that require them (e.g. defence of a poster, viva voce, and job interviews!).

10.3.1 The purpose of presentations

The assessment of presentations is usually divided into two broad areas: content, and the communication of the content (or 'substance' and 'style'). Both of these elements are important and you need to be careful not to focus on one at the expense of the other. For example, it is possible for your presentation to have relevant and informative content, but if you don't communicate it in a well-structured and engaging manner you would score relatively low marks in an assessment. Equally, you should avoid trying to dress up content that you don't understand and haven't put much effort into researching by giving a slick presentation with a few impressive slides. If you do this, it is usually fairly obvious to the audience (and it certainly will be to your lecturer) that your presentation is all style and no substance. So, both substance and style are important.

In this section, we will consider the following brief for a sample presentation:

> Prepare and deliver a 5-minute oral presentation on the use of global positioning systems (GPS) in the study of migrating populations. The presentations will take place at 12 noon on Friday 1st March in room 101.

10.3.2 Presentation structure

The starting point for selecting and organising your material should be to choose your main points. Hopefully, if you have followed the steps from previous sections of this book, you have a solid plan for the main points you need to make in your presentation and the order you want to make them in. You may well also know from analysing the brief (Chapter 7, section 7.3) how many slides you should be creating and how long your presentation is! This is essential information for creating a scientific presentation, so if you do not know these, go back and get this information.

You might not be able to list all of the main points straight away, but it is important nonetheless to make a good attempt. As you continue in your preparation you may well come back to your main points to refine or reorganise them, but you need a basic framework at an early stage to focus your thoughts and provide a structure for the rest of your material.

An example structure for a five-minute oral presentation on the use of global positioning systems in the study of migrating populations was given earlier, as follows:

- what is GPS?
- applications to migrating populations;
- examples: wildebeest and Canada geese;
- conclusion.

This is a very sketchy outline, but it is enough information to get you started. A useful technique to help you draft your main points is simply to put your research notes to one side, then try to write your main points without looking at them, expressing each point in a few words or a short sentence. Alternatively, you could try to articulate your main points to a friend, because expressing ideas out loud can be a good way of helping you decide whether or not they are making sense. It may also be helpful to add some notes to your outline to give you an idea of what the focus of your presentation should be. For instance, from the list already given, it looks like the four bullet points (or at least the first three) are equally sized sections, which probably isn't your intention. It might be better, therefore, to annotate your main points with a bit more information, as follows:

- What is GPS?—just brief background, enough for people to understand the basics.
- Applications to migrating populations—how the technology has been used in this context, and an overview of its applicability and value.
- Examples: wildebeest and Canada geese (more detailed descriptions of the application to these specific examples).
- Conclusion—summarise the main points and invite questions.

This is also the point at which you would begin to organise your main points into a logical sequence, so that each point links to and builds on the previous one. The sequence of your main points will be determined not only by what is logical, but also by what you are aiming to achieve. For example, if you are trying to build an argument you will want to move from background information to precise points of detail, or if you are explaining a process you might begin by explaining its purpose and then take your audience through each stage of the process step-by-step. So, our oral presentation on migration might go something like this:

What is GPS? GPS stands for Global Positioning System and most of you will be familiar with its application in the context of in-car navigation or the mapping software on your phone. As we can see on the second slide, the system is based on a network of satellites linked to ground stations around the world. To find where you are, the system measures the distances between your personal unit and three or four satellites, as you can see here … The exact location is then worked out by triangulation.

What you need to keep in mind is that you are trying to communicate information in manageable chunks, helping the audience to understand the progression of your argument or process. It doesn't matter at this early stage if you can't sequence your main points precisely (for instance, will it be better to deal with wildebeest or Canada geese first?), because such a level of precision is not necessary at this point. The important thing is simply to choose what your main points will be and make a good attempt at the sequencing; you can always revise it later on.

Once you know what you are going to cover, it's time to start actually putting the presentation together. This will require you to create a script of what you are going to say and some presentation aids. We are going to start with the latter, but you might wish to get your notes written first and then create aids to supplement it.

10.3.3 Designing presentation aids

This section will provide some guidance on visual design for presentations, which are similar but slightly different to the best practice for assessments more focused on a single design product (e.g. a poster or infographic).

Purpose of slides

It is our experience that most scientific presentation assignments at university will expect you to create a series of presentation slides that are submitted prior to actually delivering the presentation. There are multiple formats slides might be used as part of the assessment (check your assignment brief for details):

- On a big screen behind you as you deliver the presentation to your assessors (and possibly the rest of your cohort) in person.
- Via shared screen functions during a Teams (or Zoom, etc.) live but remote presentation session.
- As the main submission with the spoken elements submitted separately as a script or in the notes section of the file. Or with videos of you delivering the verbal part of the presentation embedded into the file.
- As the background during a recording of you delivering the verbal presentation materials, which is submitted as a video file for grading.

It is very important to remember that slides are visual aids that should supplement the points you make during your verbal presentation. They absolutely should NOT be a verbatim repetition of what you are saying.

Slide design

Slides should be used to: (a) highlight the most important words, phrases, and points you want the audience to take away from your presentation, (b) provide visual elements that make it easier to explain complex topics effectively or engage the audience in the subject matter.

Verbatim repetition of your spoken notes on a slide encourages two pieces of bad practice:

- Monotonous delivery of the verbal element that is not directed at the audience.
- Cluttered slides with too much to for the audience to read, which interferes with their ability to listen to you (Figure 10.18).

Figure 10.18 A difficult-to-work-from slide

What is GPS?
- The global positioning system (GPS) is the only fully operational global navigation satellite system
- More than 25 GPS satellites are in orbit around the Earth, transmitting signals that are picked up by GPS receivers, which, in turn, establish the receiver's location, speed and direction
- GPS is a crucial aid to navigation

Figure 10.19 An easier-to-work-from slide

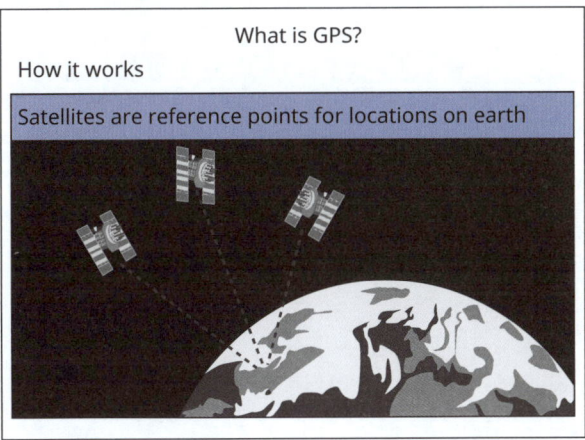

If you either condense the bullet points to keywords or phrases, or better, use an image to illustrate what you want to say (as shown in Figure 10.19), you will find it much easier to talk from. As a tip, if you do have text on your slides, avoid reading it out: if they are bullet points, expand around them; if you have used fuller text, e.g. a quotation, then allow the audience to read it, rather than reading it aloud for them.

It is worth remembering that the slides that your lecturers work from may well not follow these rules of good practice. The most common reason for this is that lecturers' slides need to be used by students for revision, and need a bit more text on than is ideal in order to be useful. More often than not, you need to use a presentation to evidence that you understand the subject now, and to be interesting and engaging in your communication of it. So, the guidance above is designed to achieve this.

It is worth visiting the guidance on visual design for bioscientists on the use of text formatting (section 10.2.3), visual elements (section 10.2.5), and colour schemes (section 10.2.6). The guidance on these elements is useful for creating well-designed and visually engaging presentation slides.

How many slides?

The answer to this will depend on the assignment brief. It may have specified the number of slides you are to use as well as the time limit, or it may have just given the time limit (as is the case with our example). If you haven't been given a specific number of slides to produce, then aiming for a slide for every 60 seconds of presentation time is a helpful rule of thumb (but you can be flexible around that). Depending on the aim of the assignment, this decision might be very easy or require a bit of refinement as you progress through the creation process. For example, if you are doing a scientific presentation that covers the results of a specific experiment (see Chapter 8 for guidance on what these sections include), then you are most likely going to have the following structure:

- Title slide
- Overview

- Introduction
- Methods
- Results
- Discussion
- Conclusion
- References/Questions

If you've only got five slides to work with, you might reduce this to:

- Title Slide
- Introduction
- Methods
- Results
- Discussion

And then have the references as footnotes on each slide.

Hopefully, you will be able to get a rough idea based on how you have done in the previous section on *Presentation structure* (10.3.2). But this might be fluid and change as you progress.

More interesting presentation aids

A key method of generating interest in what you have to say is to give the audience an opportunity to actively engage in your presentation. You may have experienced this within your lectures, where your lecturer has put activities, surveys, or quizzes into their lectures (which is called active learning), or where they have brought in an artifact for you to inspect more closely (e.g. skulls of different animals to help you understand the relationship between form and function).

It may well be worthwhile considering whether there are any activities you can include within your talk that will help evidence your understanding of the subject while actively engaging your audience. Perhaps you could bring in a GPS device to illustrate how compact they are and how easily the technology can be applied to the study of migrating populations. Consider anything that will bring what you are saying to life.

10.3.4 **Preparing presentation notes**

While the presentation aids are important, you also need to pair them with a clear, engaging, and confident spoken element. To help with this, you should really have a plan for what you are going to say. We will call these your notes, and you may well be able to use them during the presentation in order to facilitate your delivery. You have two main options available to you here: you could write them out word-for-word (known as verbatim) or you could use outline notes (e.g. bullet points). Alternatively, you could use a combination of these two types. We will consider each of these two options in turn and then look at how they can be combined to best effect.

Verbatim notes

Verbatim notes have many advantages and disadvantages.

Advantages of verbatim notes

Verbatim notes have the obvious advantage that you start the presentation knowing exactly what you intend to say. Knowing exactly what you intend to say can be reassuring and so boost your confidence. Verbatim notes can be particularly helpful, therefore, if you are lacking confidence in a particular presentation:

- perhaps because your first language isn't English;
- or you are presenting information that you are relatively unfamiliar with;
- or you are presenting complex material that requires a level of precision that would be difficult to achieve from outline notes;
- or you have some other reason for feeling nervous.

Disadvantages of verbatim notes

Verbatim notes do have significant disadvantages. There is a danger that you will simply read out your presentation and therefore won't communicate it effectively. When you look up towards the audience you might lose your place in your notes. Also, a read-out presentation is difficult to bring 'alive' because:

- you will lose eye contact with your audience;
- your voice will probably become more monotonous;
- you will have much less energy and enthusiasm about the delivery;
- as a result, your presentation will be less engaging.

How to use verbatim notes

Verbatim notes seem to have got themselves a bad name. Often people will tell you not to use them because of the very real disadvantages listed previously, but this advice usually comes from people who have become experienced presenters and so have forgotten about much of the anxiety that is often associated with presenting. However, as can be seen from the advantages, there are very real benefits, too. The trick is to use verbatim notes in the right way. We will take each of the potential disadvantages in turn and suggest a strategy to help you avoid that particular pitfall.

There is a potential danger that you will simply read out your presentation

Don't! Reading out the presentation will nearly always result in the problems listed. You don't need to memorise your notes, but you do need to have practised them sufficiently so that you can work from them as a prompt, as opposed to as a script.

If you look up from your notes you might lose your place

Quite possibly. So, make sure you format your notes in such a way so as to help you not lose your place. For instance, increase the font size, increase the line spacing, use paragraph breaks, use bullet points, use annotations, and highlight keywords. Figure 10.20 shows verbatim notes

Figure 10.20 Unhelpfully formatted verbatim notes

> **The use of GPS in the study of migrating populations**
> Hi, my name's Sam and my presentation is about how GPS helps scientists to study migrating populations.
>
> This presentation is divided into three main sections, as shown on the first slide. First, I'll give a brief explanation of what GPS is. Then I'll explain how the technology has been applied to the study of migrating populations. Finally, I'll tell you about two specific examples: wildebeest and Canada geese.
>
> First then, an explanation of what is GPS? GPS stands for Global Positioning System and, as its name suggests, it forms a worldwide navigation system meaning you should never get lost again! As we can see on the second slide, the system is based on a network of satellites linked to ground stations around the world. To find where you are, the system measures the distances between your personal unit and three or four satellites, as you can see here . . . The exact location is then worked out by triangulation (see, you always knew that trigonometry you slogged through at school would come in handy).
>
> The first GPS systems were developed in the late 1970s for the United States military. Since then the technology has developed considerably and in the 1980s became available for civilian use. GPS is now used for many civilian purposes, including: map making, land surveying, and of course, navigation. The particular use we are interested in, however, is in its application to the study of migrating populations of animals. Studying migrating populations using GPS has only been possible since the technology required to track animals has reduced in size. The necessary tracking tags are now small enough to track many different sorts of animal, not just large ones. Tags then relay information, via the GPS satellite system, to monitoring units, thus allowing scientists to know the precise coordinates of the animals they are studying. Such detailed information is invaluable for scientists, as reviously they were simply not able to know about an animal's whereabouts in such detail.

that would be very easy for you to lose your place with; Figure 10.21 shows more helpfully formatted verbatim notes that dramatically reduce the chances of you losing your place.

A read-out presentation is difficult to bring 'alive'

This is because you will lose eye contact with your audience, your voice will probably become more monotonous, you will have much less energy and enthusiasm about the delivery, and as a result, your presentation will be less engaging.

Again, most of this will be helped by practice. If you can get to the stage where you have practised your presentation sufficiently so that your notes are a prompt, rather than a script, you should be able to think about the communication, as well as content issues—issues such as making eye contact with the audience, varying the tone of your voice appropriately, and generally trying to be more engaging.

Outline notes

Outline notes also have advantages and disadvantages.

Figure 10.21 Helpfully formatted verbatim notes

The use of GPS in the study of migrating populations
Don't start until you're ready

Introduction

Hi, my name's Sam and my presentation is about how GPS helps scientists to study migrating populations.

This presentation is divided into three main sections, as shown on the first slide.
1. First, I'll give a brief explanation of what GPS is.
2. Then I'll explain how the technology has been applied to the study of migrating populations.
3. Finally, I'll tell you about two specific examples: wildebeest and Canada geese.

outline slide

1 min. max.

What is GPS?

First then, an explanation of what GPS is? GPS stands for Global Positioning System and, as its name suggests, it forms a worldwide navigation system meaning you should never get lost again!

As we can see on the second slide, the system is based on a network of satellites linked to ground stations around the world. To find where you are, the system measures the distances between your personal unit and three or four satellites, as you can see here . . .

Advantages of outline notes

Having your notes in only outline form can free you up to focus on engaging with the audience, rather than focusing on your notes. Also, if you are not reading your notes out, your voice and manner will probably be more natural, which will also help engage your audience. There is no temptation to read your notes out verbatim, which, as we highlighted earlier, causes problems.

Disadvantages of outline notes

There are two main disadvantages of outline notes. First, you might be more nervous about your presentation because you have fewer notes to rely on. Secondly, you may find that your mind 'goes blank' during your presentation because your outline notes aren't a sufficient prompt to remind you of what you intended to say.

How to use outline notes

As we have seen with verbatim notes, outline notes have both advantages and disadvantages; the trick is to use outline notes in the right way. The only way around these

Figure 10.22 Outline notes for the introduction

Introduction
- Me
- Title of presentation
- What I'll cover (slide 1)

potential disadvantages is practice: you need to rehearse your presentation a sufficient number of times so that you know your outline notes are full enough to remind you of what you need to say. Of course, while you are testing the adequacy of your notes in this way you will also be learning the material. Remember though, just because you can get your presentation right once or twice in the privacy of your bedroom doesn't mean that everything will be fine in the more pressured situation of a tutorial class or assessed presentation.

Combine note types to maximise benefits

Given the potential disadvantages of each technique, probably the best method to employ is a mixture of the two. Where you are confident that you only need short bullet points to prompt you, you can use outline notes, for instance, for the introduction, as shown in Figure 10.22. However, where the material is more complex and precision is more important, you can use at least fuller, if not verbatim, notes, as shown in Figure 10.21.

Openers and closers

In addition to planning the main content, it is also important to think about how you will open the presentation and how you will close it. It is a good idea, regardless of what type of notes you choose to use, to know exactly what your opening statement is going to be (the 'opener') and exactly what your closing statement is going to be (the 'closer'). If you have a faltering start, perhaps tripping over your opening words or not saying them clearly, it is likely that you will lose confidence and composure, which can be very difficult to regain. Opening statements need only be simple sentences such as:

'Hi, my name's Sam and my presentation is about how GPS helps scientists to study migrating populations.'
   'Hi, my name is Alex and I've got five minutes to talk to you about ...'

Simple statements such as these get you off to a positive start and make it clear to the audience what you are doing. Equally, closing statements are important, too. Too often, presentations finish with a mumbled apology: 'Err, that's it, I've finished …', then the pre-senter sits down. A much better way to finish is a clear conclusion followed by a closing statement such as:

> 'So, in this presentation we have seen how GPS can be used as a very effective tool for monitoring the migration of very different types of animal, but also that there are some specific limitations that need to be addressed.
>
> Thank you very much for your attention and I will be happy to try to answer any questions.'

10.3.5 Practice

We need to highlight the importance of practice as, often, how effective a presentation is will depend to a large extent on how much you have practised it. Practising a presentation is important because if you know your presentation well it frees you up to think about the communication, as well as content issues (style, as well as substance).

We noted in *How to use outline notes* in the previous section, that just because you can get your presentation right once or twice in the privacy of your bedroom, it doesn't mean that everything will be fine in front of an audience. However, the value of practising in your own environment is not to be underestimated, and there are several ways you can make it more effective.

Practise against the clock

It can be very difficult to judge how much time your presentation will take without actually practising it out loud and against the clock. Depending on the style of notes you have used, sometimes you will look at your notes and think that it will barely last 2 minutes, never mind 5. Other times you will think you have got far too much material, but in reality, you will get through it far quicker than you thought. The only way to have a reasonably accurate idea of how long your presentation will take is to practise it out loud. Of course, by doing this, you are not only getting an idea of how long it takes, you are also learning the material and refining its delivery.

Practise in front of a mirror

Just as it is difficult to know how long the presentation will take, it is also difficult to know how you will look. The only way to get an idea of this is to practise the presentation in front of a mirror or, better still, record yourself on your phone. You will probably feel a bit self-conscious doing this, but it will give you some very useful feedback about how you are pre-senting yourself, as well as how you are presenting your material. It is a particularly helpful way of assessing whether or not you are using your body language appropriately (see *Think about your body language* in section 10.3.8).

Practise in front of your friends

Most usefully, you could practise in front of some willing friends. This could give you feedback on your presentation's length, your body language, the structure, in fact all aspects of your presentation. It can be a bit embarrassing and you will doubtless make mistakes, but as long as your friends are going to give some constructive comments, and you can deal with the potential embarrassment, it will give you invaluable feedback and so improve your final presentation, and thereby its grade, significantly.

10.3.6 Common concerns with public speaking and how to address them

The differences between written and spoken presentations are often the cause of concern or anxiety about delivering a presentation. For example:

- What if I'm so nervous I forget what I was going to say?
- What if the technology doesn't work?
- What if I lose my place in my notes?
- What if the audience doesn't seem to be following what I'm saying?
- What if I get asked questions that I don't know the answers to?

These, and other concerns, are important things to consider. While listing possible concerns in this way can seem a little overwhelming, it is important to identify them so that something can be done about them. You might be able to relate to all of these concerns, or perhaps to just a few of them, but it is likely that you will experience all of them at some point or other, so it will be useful to address each of them in turn.

Deal with anxiety

It's all very well saying 'deal with anxiety'; it is much more difficult to actually do it. It is also quite difficult to write about how to deal with anxiety, because different things make different people anxious. However, for the purposes of this chapter we are going to assume that standing up in front of a group of people to give a presentation is something that causes many people to be anxious or nervous. If it doesn't, it probably should!

As with Chapter 6, section 6.2.3 on stress levels in exams, we have deliberately titled this section 'Deal with anxiety', rather than, for example, 'Get rid of anxiety' for a couple of reasons. First, you can't get rid of all your anxiety, so it would be unrealistic to try; and secondly, anxiety or nervousness can be helpful, so even if you could, you don't want to get rid of it altogether. Anxiety can motivate you to do the work needed, so that you are well prepared for the presentation, and it can also make you more alert and energised during the presentation. However, if your anxiety becomes too pronounced, difficulties can occur, which may impair your ability to prepare effectively for, and perform during, the presentation.

Table 10.2 Replacing negative thoughts with positive ones

Negative thought	Positive replacement
'It will be a disaster.'	'I will aim to do the best I can.'
'I never do any good at this kind of thing, it's bound to go horribly wrong.'	'Just because I had a problem with this is in the past does not mean that things are bound to go wrong.'
'I will fail my degree and never get the career of my choice if I don't do well in this presentation.'	'The marks for this presentation are only a small percentage of my overall degree. If I don't do as well as I would like there will be other opportunities to improve my marks.'

A useful strategy in dealing with anxiety is to identify what your concerns are so that you can address them, which is what this section is about. In general terms, however, the following strategies can be helpful:

- try to replace negative thoughts with positive ones (Table 10.2);
- practise steady, deep breathing before and during your presentation;
- have a bottle of water to hand—it will help stop your throat drying out, but better still, if your mind goes blank you can simply stop and take a sip, which usually gives you just enough time to gather your thoughts and regain your composure without the audience noticing.

Try this—A one-minute presentation

Get together with a small group of friends. Each choose a topic with which you are comfortable, for example, your favourite hobby, holiday destination, or similar topic with which you are familiar. Now, prepare a very short presentation on this topic. Each of you should now present the talk to the rest of the group. The talk should be no longer than one minute. One member of the group should keep time.

Have a backup plan

The second concern that we identified was 'what if the technology doesn't work?' This is a legitimate concern and one that requires consideration. Broadly speaking there are two possible causes of technology apparently not working: first, it could be a genuine fault with the technology; or secondly (and perhaps more commonly), it could be a lack of competence on the part of the person operating it. Both types of problem can be minimised by having a backup plan, but be careful not to jump too quickly to using your backup plan, when it could be, for instance, that you are just pressing the wrong button!

It is important, therefore, that you familiarise yourself with any technology that you intend to use. Amid the pressure and stress of a presentation about to start, or already started, it can be easy to panic and not think straight. Make sure you understand how to use what you are using. If possible, practise not just with something like it, but with the actual technology in the actual room where you will actually be presenting. The closer you can get to practising with the real thing, the more confident and prepared you will be when it comes to the presentation.

One common source of technological problems is the transfer of the electronic file to the computer being used to deliver the presentation, for example, because of failure of a memory stick. If you have the chance to access the venue before your talk, it is always a good idea to copy the talk across onto the desktop of the computer and check it is working. If you can't do that, then having the talk on a backup medium is also a good idea, for example, have a spare memory stick, or have the talk emailed to yourself so you can access that way.

For those occasions when the problem is with the technology, rather than the person operating it, this is where you will have to use your backup plan. Some forms of technology are less reliable than others, and the more multifaceted the output the more potential there is for it to go wrong. For instance, if you are using slides, video, audio, and a live internet connection, there are a lot of things that can potentially go wrong! You need to consider whether or not such a range of technology is necessary to your presentation. There would be much less risk if you used less technology and it would be much simpler. Indeed, it may even improve your presentation (more technology does not necessarily mean better presentations).

Make your notes work for you

The third concern that we identified was 'what if I lose my place in my notes?' As we pointed out in section 10.3.4 *Preparing presentation notes*, there is a lot you can do to help yourself here. First, make sure you use an appropriate type of notes: outline or verbatim (or a combination of both). Secondly, make sure you format them in a way that makes them easy to use and so it is less likely that you will either lose your place (a possible problem with verbatim notes) or forget what you were going to say (a possible problem with outline notes). Section 10.3.4 also highlighted the benefits of combining these note types to maximise the benefits, using outline notes where you don't need much prompting and verbatim notes where you do.

Help the audience follow what you are saying

The fourth concern was 'what if the audience doesn't seem to be following what I'm saying?' This is clearly an important consideration for the benefit of the audience, but it is also a significant factor in your own performance as a presenter. If you begin to notice, during your presentation, that your audience is distracted and perhaps seemingly confused, it can be very difficult not to let that influence your confidence and you can begin to lose your composure as a result. Obviously, it is important not to be over-sensitive here; it can be easy to interpret normal audience behaviour, such as someone yawning, as a signal that your presentation is going really badly, when this probably is not the case. It is likely that there will always be some people in the audience who have had a bad night's sleep or are thinking about what they are going to do at the weekend, or perhaps just exhibiting body language and facial expressions that are out of sync with their genuine response. Don't let one or two negative responses unduly influence your confidence. Instead, there are several things you can do during your presentation to help your audience follow what you're saying:

- use verbal signposts;
- use your voice;
- be aware of your audience's needs.

Let's consider each of these in turn.

Use verbal signposts

If a presentation is well-structured then the audience can see the direction the presentation is taking and so should be able to follow it more easily. We said that one of the ways you can do this is by letting the audience know what the structure will be; both at the beginning and as you go along. For example, at the beginning of a presentation you should map out where you are going by telling the audience what you will cover. Letting the audience know what the structure will be may involve you saying something like this (as seen in section 10.3.2 *Presentation structure*):

> 'This presentation is divided into three main sections. First, I'll give a brief explanation of what GPS is. Then I'll explain how the technology has been applied to the study of migrating populations. Finally, I'll tell you about two specific examples: wildebeest and Canada geese.'

In addition to mapping out the structure at the beginning of the presentation, it is also helpful to give the audience prompts as you go along. If you were reading a written presentation, you could simply flick through the pages, and read the headings and subheadings, but clearly you can't do this with a spoken presentation. So, help your audience follow what you are saying by telling them what the structure will be and then use pointers as you go along to make it clear where you are up to, for example:

- a single word such as 'First', 'Secondly', and so on;
- linking statements such as 'So that's the background to the technology, let's now think about its application . . .';
- concluding statements such as 'In conclusion . . .' or 'Finally . . .'.

Use your voice

Use your voice to help your audience follow what you are saying. You can use it in many different ways by varying the volume, pace, and pitch.

Volume

Make sure that your voice is loud enough for your audience to hear clearly (it's surprising how many presenters don't do this—nervousness will often make you speak more quietly). Speaking too quietly (or too loudly) can make it difficult for your audience to follow your presentation. In normal conversation, people tend to raise or lower their volume for emphasis. For example, a person may speak loudly when giving an instruction, but softly when apologising. Try to use these normal conversational variations in your own presentation (but don't go too quietly!)

Pace

Make sure that the speed of your delivery is easy to follow. If you speak too quickly (again, this is often caused by nervousness) or too slowly your audience will have difficulty in following your talk. To add life to your presentation, try changing the pace of your delivery. A slightly faster section might convey enthusiasm. A slightly slower one might add emphasis or caution. As you become more experienced, this kind of variation will happen much more naturally, as it does in conversation (especially as you become better at dealing with the anxiety), but it's a good idea to be deliberate about it while it doesn't come naturally to you.

Pitch

The pitch of your voice also varies in day-to-day conversation and it is important to play on this when making a presentation. For example, your pitch will rise when asking a question; it will lower when you wish to sound severe.

Be aware of your audience's needs

If you are giving a 5- or 10-minute presentation to a group of your peers and a tutor, then you will be limited as to how much you can actually do about your audience's needs. You will probably be on quite a tight schedule (often you will lose marks for running over time) and if there are several of you presenting, then it will often be a case of just making sure the presentations fit into the allotted session. Additionally, when you are relatively new to presenting, you will want to concentrate on just delivering what you have prepared, rather than making any changes to your presentation as you go along in response to the reaction you are receiving. However, if you are at least halfway down a list of, say, six presentations that are scheduled for a session, there is one simple thing you can do (if your tutor doesn't suggest it)—simply suggest a brief break. The break does not have to be long and it will probably be best if people don't even leave the room. However, if you suggest a break of a minute or so, while you are setting up, perhaps even suggesting that people stand up and walk around a bit, it should be just enough for people to wake up and re-energise themselves so that, when you start your presentation, they are that bit more attentive, making it easier for both you and them.

10.3.7 Respond to questions appropriately

The final concern we identified was 'what if I get asked questions that I don't know the answers to?' This is a common concern, because it is the bit of the presentation that you have least control over and it can be difficult to know what kind of questions you might be asked. If you are doing an assessed presentation, one important piece of information to find out is whether or not you are required to invite questions, and if so, whether or not your responses count towards the overall mark you receive. Whether or not the questions (or more accurately your answers!) count towards your mark, there are several things you can do to make this part easier.

Create a list of potential questions and prepare answers

We said earlier that it can be difficult to know what kind of questions you might be asked, but it is not impossible. In fact, with just a bit of thought you can probably come up with several potential questions that you might get asked. Questions will usually focus on areas of the presentation that the audience found particularly interesting, or perhaps a section that needed more explanation or additional context. Obviously, you can't predict all the questions you might be asked, but you will be able to predict some, and that will help to take some of the pressure off.

Try this—Identifying possible questions

Show a draft of your slides and presentation notes to a friend or peer. Ask them to come up with two or three questions that occur to them while they are reading your material. Draft an answer to each question and get feedback on the helpfulness of your answer from your colleague.

Say whether you will be taking questions and when

It is a good idea to state clearly at the beginning of your presentation whether you will be taking questions and, if so, when. That way the audience is clear about if and when they can ask questions, but also it means that you are more in control of the situation, which is important—lack of control, as identified earlier, can make the questions part of a presentation difficult. Sometimes people like to deal with questions as they go along as it can allow more interaction. However, it is usually easier to deal with questions at the end so you can deliver your talk free from interruptions, which you might find distracting. Taking questions at the end also means that you keep control of the timing of the talk, which you may find difficult to do if there are questions being asked as you go along.

Ask for clarification

When you get asked a question it is often unclear what the question actually means. If you are going to answer the question that the questioner wants to ask, rather than the one you think they might be asking, then it is important to ask for clarification. For instance, you might get asked the question:

> Do all the animals follow the same migrating pattern year after year?

A good way of clarifying the question is to rephrase it and then ask if that is what was meant. For example:

> When you say 'all the animals' do you mean you mean all migrating populations of wildebeest or all migrating populations of Canada geese?

Clarifying questions in this way is useful for a number of reasons:

- it gives the questioner a chance to refine the question;
- it ensures that the rest of the audience has heard and understood the question;
- it enables you to check that you have understood the question;
- it also gives you thinking time to prepare a better answer.

Answer the question

Preparing potential questions and answers, saying if and when you will take questions, and asking for clarification will all help you to answer questions more appropriately. However, when you actually open your mouth to answer the question make sure you:

- answer the question you have been asked, rather than the one you want to answer (seeking clarification helps here);
- are focused and to the point—answer the question and then stop, don't start waffling;
- are honest about the limitations of your knowledge—say if you don't know the answer (it will usually be obvious if you are bluffing);
- say if a question is beyond the scope of the presentation—e.g. 'That's a good question but my research for this presentation focused on tracking migrating populations by GPS, not by any other means'.

10.3.8 Some key techniques to master

Finally, it's important to think about some key techniques that you need to develop if you are going to deliver your presentation effectively.

Relax!

The first key technique is to try to relax. Clearly, this is easy to say, but much more difficult actually to do. However, if you can relax, even just a little bit, it will help your presentation a great deal. Presenters who are tense will deliver their presentation in a stilted, unnatural, and nervous manner, whereas a presenter who is relaxed (or slightly less tense at least) will be more natural and more confident in their delivery. We addressed many of the issues that will help you to be more relaxed in *Deal with anxiety* in section 10.3.6, but we also wanted to highlight it here as a key technique because it makes such a positive difference to a presentation.

Be conversational

In section 10.3.4 *Preparing presentation notes*, we highlighted the importance of getting to the stage where you have practised your presentation sufficiently so that your notes (whatever their format) act as a prompt, rather than as a script, thus freeing you up to think about the communication, as well as content issues. Something to aim for is to try to be conversational. Being conversational doesn't necessarily mean being informal or chatty (this is often not appropriate in an academic context), but it does mean that you are speaking in a natural way, as if you were actually having a conversation. Clearly, a five-minute presentation is a fairly one-way conversation (!), but if you can aim to express your-self in normal, everyday (though still appropriate) language, it will help the presentation to be more engaging, and being engaging was one of the key characteristics of effective presentations that we mentioned in section 10.3.4. It will also help with using your voice (volume, pace, and pitch) as we mentioned in *Help the audience follow what you are saying* in section 10.3.6.

Think about your body language

Body language is an important and often neglected element of a presentation. It includes facial expressions and eye contact, as well as what you do with your hands, and where and how you situate yourself in the room. Body language can be both positive and negative—it can help a presentation be more engaging, or it can distract from or even counteract the presentation's content. If you have practised your presentation in front of a mirror, recorded yourself on your phone, or presented to your friends (as suggested in section 10.3.5) you will be more aware of how your body language influences the way you present. Even though it can be counterproductive to become too self-conscious of your body language, here are some suggestions to make sure you use your body language positively.

Think about how you will use the available space

Where are you going to stand in the room (you might prefer to sit, but it's more normal to stand and it helps you project your voice better)? Where will you put your notes? If you are using a projection screen (with an overhead projector, slides, or Prezi), you need to be able to stand somewhere that doesn't obscure anyone's view of the screen, but still allows you to be close enough to move to the next element without getting in anyone's way or tripping over cables. Spend some time thinking about this before you start presenting because once you have started your presentation you want to be able to concentrate on what you are say-ing, rather than thinking about where the best place is for you to stand.

Think about what you will do with your hands

People use their arms and hands in everyday conversation to add emphasis or to help describe events. Presenters will therefore look rather awkward if they keep their hands in their pockets or rooted firmly at their sides. Equally, you don't want your hands to move in a way that distracts your audience's attention such as nervous, repetitive movements. Instead, use your hands to emphasise and enhance what you want to say verbally. As long as use of your hands is controlled and purposeful, then it has a key role to play in support-ing your verbal communication. If you can try to relax a little and be conversational then it is more likely that what you do with your hands will be more natural and more in sync with what you are saying.

Make eye contact

Often during an assessed presentation, the presenter only makes eye contact with the person doing the assessing, thereby excluding everybody else. If you were to have a conversation with someone, you would make eye contact with them, and not doing so would probably be considered a bit rude. So, when presenting, try to make eye contact with everyone. It is likely that you will only be presenting to a relatively small number of people, so making eye contact with everyone is a realistic aim. If, however, you are presenting to larger groups just try to make eye contact with all parts of the room. Making eye contact can be more difficult if you are using verbatim notes, so make sure you follow the guidelines on how to use verbatim or outline notes in 10.3.4 *Preparing presentation notes*.

Smile

If you are nervous, it can be difficult to smile, but smiling can make a big difference to how people react to you. Clearly, you don't want to have an inane grin on your face for the entire presentation—that would be unnatural and look ridiculous—but making a point of trying to smile, at least before you start, will help your audience feel more at ease. It is likely that at least someone in the audience might smile back at you, which in turn helps you feel more at ease, too.

Show some enthusiasm!

Finally, show some enthusiasm! If you appear bored by, and uninterested in, what you are saying, your audience will probably react in the same manner. Conversely, if you can show genuine interest and enthusiasm, it is more likely that your audience will be willing to listen attentively and work at following what you are saying.

Try this—Practising your presentation

Get together with a small group of other students before you all have a presentation due. Each student should give their presentation, ideally in the room that you will be using. For each presentation, each member of the audience should write down one example of good practice and one area needed for improvement, and provide this written feedback to each presenter at the end of the session.

 ### Chapter summary

Contemporary Bioscience degrees expect students to develop a broad range of communication skills beyond the traditional long-form written essay. Particularly important are the visual and oral skills required for graduates to communicate clearly, concisely, and engagingly to a variety of different audiences and digitally enhanced formats. Section 10.2 of this chapter has provided guidance on how to use visual design skills to produce products that effectively blend written and visual elements to communicate ideas (such as posters and infographics). While section 10.3 focused on coursework types that require the use of written, visual, and oral communication skills (such as presentations). For some of you, these types of assignments will feel like a great opportunity for some creativity and freedom, while for others, these will be highly anxiety inducing experiences. Hopefully, this chapter has provided some guidance that will be useful to both of these people!

11 Revising drafts and finishing touches

Introduction

If you have been following along with the structure of this book, you should now have a finished draft of whatever assignment you have been creating. It should be well-researched, well-planned, structured according to the key points required to meet the assignment brief, and be a good example of the communication format (written, oral, or visual). It may well feel 'complete'! However, your first full draft of any assignment is unlikely to be the final thing that you submit (as long as you have left enough time for creating it). Reviewing and improving are important stages of the whole production process, but they often get overlooked because not enough time has been left to do them effectively. Indeed, review is most effectively done when you have left a gap of a few days between writing the draft and undertaking the review. It is difficult to look at a product critically immediately after you have finished.

Most professional creative processes require an iterative cycle of creation, review, and revision. A key aspect that separates good work from bad work is the thoughtful improvement of a product over time. There is also great benefit in taking the time to proofread your work for small errors that would reduce the professionalism of the final product. This chapter provides guidance that may help you with two aspects of the iterative creation process: section 11.1 on review and improvement of a full draft, and section 11.2 on proofreading and other finishing touches.

11.1 Improving drafts

It can be very tempting to look at your first draft and say 'Yep, that will do, this has taken ages and I do not want to do anymore'. Alternatively, some people spend far too long trying to make their assignment submission 'perfect', which is impossible. To address both issues, it is important that you are strategic and purposeful with your editing. In short, as with many things, you need to have some idea of what you are trying to accomplish. In this section, we will try to outline a review process that you can use on any piece of work at any stage of study.

When planning your timeline for preparing the assignment, try to make sure that you factor in time for the review process. Ideally, you should aim to leave a gap between completing the work and reviewing it so that you can come to it with 'fresh eyes'. This makes it much easier to be critical when reading through your work.

To help with this, we have split the process into two scales: the macro-level (reviewing the product as a whole), and the micro-level (reviewing the product on a line-by-line basis).

11.1.1 Reviewing the product as a whole

Before you start to review and draft each little bit of your assignment at the level of the sentence, there are some big picture things that you need to make sure of first.

Is it under the indicative word count or content limit?

Make sure you know how your course deals with submissions that are over/under the word count as identified in the assignment briefing. We use the phrase 'word count' but this could equally apply to the number of slides or time available for a presentation, or the size of a poster. As a rule of thumb, if you are significantly under the word count, then the submission is unlikely to be providing the expected level of detail on the subject. If you find yourself significantly under the word count, then . . .

. . . go back to the assignment brief and check you have included all the expected content.

. . . go back to the research section; are there key theories you have not considered or more evidence available to support your key points?

Being over the content limit is likely to be addressed by many of the following questions, so we will leave that for now.

Does it meet the assignment brief?

Right at the beginning of this process, in Chapter 7, we provided some guidance on how to effectively analyse an assignment briefing. It can often be helpful, with fresh eyes, to review your assignment against the initial brief and you check that you have:

- answered the question/discussed the right topic;
- adhered to all rules and standards outlined in the brief (e.g. it may have specified how many figures you should include);
- included all expected contents and sections requested;
- met the standards set out in the marking criteria for the grade you hope to obtain (e.g. if it asks for critical thinking, whether you have evidenced it sufficiently—see Chapter 12, section 12.2).

Does the content provide a logical flow of information?

This is likely to mean slightly different things for different assignments, but there are some general ideas that you can apply to most assignments:

- Is the text grouped together by the subject being covered (either as a paragraph, textbox, or slide)?
- As you read the assignment, does the order of subjects make sense to you (sometimes subheadings can help the reader make sense of this)?
- Do you start off considering broad subjects that set the context for what you are going to discuss, and become narrower and more specific to the assignment title as you progress?

- Do you end with some clear conclusions that are well supported by the information that precedes them?
- Are all your diagrams and tables close to, but after, the point that they are mentioned in the text?

Does it build on feedback from previous assignments?

Using feedback is a very important part of the learning process and is covered in detail in Chapter 13 *Making the most of feedback*. However, it is worth highlighting some points on this here. Feedback on assessment should come in three forms:

1. Subject-specific feedback, for example, the identification of factual errors or omissions.
2. Format-specific feedback related to the chosen form of communication used in any given assignment (e.g. feedback on your oral delivery of a presentation).
3. Generic feedback that applies to most assignments (e.g. corrections on reference formatting or guidance on critical thinking).

It is very important that you review feedback you have received on other assignments that share a format with the one you are doing now. These are likely to include guidance on how to communicate more effectively, or help you understand the expectations of the grader. This is often a great time to review this feedback because you are likely to be less emotional and have a specific product that you are reviewing (as opposed to a hypothetical future assignment).

11.1.2 **Reviewing the content on a line-by-line basis**

In addition to spending time reviewing the submission as a whole, you might also benefit from reviewing the content at a much finer scale. A common issue with student-written submissions is that they include content mainly to bring a product up to the word count, with much of the content being repetitive, long-winded, or irrelevant. For example, one commonly sees posters with massive pictures of the study species on them, but the student hasn't really considered whether that is the best use of available space (it might have been better as a graph because the assignment was really about data presentation). On the other side of the spectrum, some students try to cram too much content in because they have found lots of information at the research stage but have not been sufficiently critical about whether that research is relevant or appropriate. With this in mind, it is useful is to consider whether a sentence, paragraph, or other content element (such as a figure) is making the best use of your limited word count. We have included some questions to ask as you review your assignment at this level of detail.

Is this sentence relevant to the aims of the assignment?

When deciding whether information should be left in or taken out, ask yourself this question:

'Is the information that I'm thinking of including going to make what I am trying to say any clearer?'

Of course, this relies on you knowing what it is you are trying to say, which is why establishing the main points first is important.

Let's consider a hypothetical presentation on the value of GPS technology for our understanding of ecological systems. In your research, you may have found out lots of information about the global positioning system, including that the first experimental satellite for GPS was launched in 1978, but will this piece of information help to clarify what you are trying to say? If you have decided that the focus of your presentation is the example case studies that provide examples of specific changes in our understanding of ecology (e.g. differences in migration behaviours between wildebeest and Canada geese), then you will probably decide that incidental detail about the history of GPS is unnecessary.

When reviewing your work, you should always come back to the bigger, more fundamental question, 'is this relevant?' Also remember that when you decide to leave information out, it doesn't necessarily mean that the information is wasted: the information you leave out may have been helpful in leading you to other information. It may also be useful supplementary information if you are asked questions about your presentation, or it may be useful if you subsequently have to write an associated assignment that expands on the one you are currently working on.

Is this point made concisely and without repetition?

More is not always better . . .

More content means the grader and/or reader must read more to find each point. Overly, long writing means you can make fewer points within the word count.

- Make your points as clearly and concisely as possible (see Chapter 10, section 10.2.4).
- Use sensible paragraph structures that guide the reader to the key point being made (see Chapter 8, section 8.1.4).
- Check that you are not making the same point in a different way across multiple sentences.

A common source of unnecessary words are sentences that say that there is a point, as opposed to merely stating the point. For example, let's consider the following paragraph which we might use to introduce an essay . . .

> There are many factors that contribute to the degradation of coral habitats worldwide. No single driver of coral habitat loss is solely responsible but some are more important than others. Research shows that the most important drivers are ocean acidification, global warming, overfishing, and water pollution. However, the relative impact of these drivers depends upon the coral reef location and El Nino events. This essay will consider the evidence for the key drivers of coral reef decline and present a plan on how they might best be addressed.

This is not necessarily poor writing, but the opening section could be much more concise because it contains sentences that state that there is a point to be made, instead of just stating the point. We could rewrite this as . . .

Research shows that the most important drivers of the global decline of coral reef habitats are ocean acidification, global warming, overfishing, and water pollution. However, no single driver is most important globally as the relative impact depends upon the coral reef location and El Nino events. This essay will consider the evidence for the key drivers of coral reef decline and present a plan on how they might best be addressed.

By removing the content that says there is a point to be made (i.e. the first sentence) and reducing the repetition of points (i.e. about no single driver being the most important), we have reduced the text from 88 to 71 words; a saving of ~19%. That may not sound like a lot but if you apply that approach to the whole of a 2500-word essay, you could save over 400 words, which is potentially a whole additional paragraph of content you could use to explain the topic in more detail!

Have I supported this point with reference to appropriate scientific literature?

The frequency of referencing is somewhat linked to the level of study, but it is expected that you reference where information has come from. If you find large sections of work with no citations in it, then it might be a good idea to go back to the research stage to shore this section up. Realistically, by the end of your degree, most sentences in an assignment should be supported by an appropriate reference.

Alternatively, you may find that some of your key points are supported by inappropriate sources. You might ask yourself 'Is the information that I'm thinking of including going to make what I am trying to say any more authoritative?' The authoritativeness of information you use will depend largely on the source of that information; whether it is well-researched, impartial, and up to date. This is usually easier to establish for publications such as textbooks and articles published in research journals than for web resources, especially freely editable sources such as Wikipedia. You could improve your content by making sure it is supported by evidence from more authoritative sources, recent and relevant scientific papers are the gold standard here (see Chapter 12, section 12.2 for guidance on reading and critiquing scientific literature).

11.1.3 When to finish the review process

The review stage is important, and you must build time into your assignment creation process for it. However, the amount of time it takes is different for each person and possibly each assignment. Some quick points to help you know how long to spend on the review and redrafting stage:

- This stage will be longer if you are the kind of person who prefers to just get something 'on to paper' as quickly as possible.
- You need to review what you have produced against the assignment briefing and on a line-by-line basis to ensure that each section is relevant, concise, and supported by appropriate citations.

- It is not possible to create something that is perfect. Sometimes the best way to deal with the review process is to allow a certain amount of time, and say it is finished when that time is up.

11.2 Proofreading the document

It is very tempting, when you have finished reviewing and redrafting the last word of an assignment, just to submit it. However, there is still more work to done. It is very important that you proofread the essay, to find any errors that will reduce the professionalism of your work. One of the hardest parts of proofreading is that, when we read through something we have written, we read what we expect to see and can very easily overlook mistakes. This is especially true if you have just finished writing the essay and are still feeling totally immersed in the exercise. If you have managed your time carefully, you will have at least a few days spare before you must submit the assignment. So, put the submission away for a few days, then sit down and read it carefully, as if it had been written by someone else and you were tasked with finding errors.

11.2.1 Is proofreading really that important?

Yes it is, for a number of reasons:

- the person who is marking your assignment will find it difficult to read text that is littered with errors and, as a consequence, you will probably get lower marks than the content alone might merit;
- some mark schemes include a specific allocation for meeting the expectations of professional written communication, such as grammar and spelling;
- errors in punctuation and syntax can lead to ambiguities of meaning, so the sense of what you are writing is no longer clear;
- unprofessional formatting (e.g. poorly aligned content on a poster) may suggest to the grader/reader that the product is rushed, unfinished, or lazily produced;
- it is important to get into the habit of writing accurately: if you are asked by an employer to write a technical report, that employer will not be pleased if there are numerous mistakes;
- furthermore, when you are trying to sell yourself in job applications, these mistakes can be very costly: employers will often filter out applicants for poor spelling or grammar.

So, get into good habits early and always check what you have written.

11.2.2 Professional expectations (e.g. grammar, etc.)

There are certain expectations of professional work that apply universally to professional environments, such as the use of correct grammar and spelling. This section will outline some simple methods of checking your work for these issues and provides a short glossary guiding usage of some common grammatical constructs.

How to proofread your work

1. Use a grammar checking tool

Many people struggle with grammar and spelling, maybe they have a diagnosis of dyslexia, or other specific learning difficulty, maybe they don't. Fortunately, most word processing software has in built spelling and grammar checkers that can catch some of the more obvious errors. These checkers are becoming more advanced all the time. At the time of writing, the online version of Microsoft word can cross-reference spellings of highly technical scientific words such as 'angiotensin' and 'osteoblast' (and tell you what they mean with an inbuilt dictionary). It will also make suggestions to help you make your writing more concise. There is even access to more advanced AI-based writing assistance programmes (such as Grammarly), which can be particularly useful if you are not confident with written English (e.g. because it is not your first language). As long as you are using these tools to revise your own content (and not rephrasing someone else's), then these tools should be viewed as helpful support in checking your work for grammar and spelling errors.

2. Read it out loud

There is a danger that writers become too reliant on the spelling and grammar checkers in word-processing packages and assume that because the program has not highlighted a piece of text, then it is correct. These packages are certainly helpful, but they are not infallible. Consequently, your essay on screen may be highlighted in lots of places and so there is a tendency to ignore all these as 'false positives'. Also, the recommendations are based on application of general rules, and may not always convey the exact meaning you wish to communicate in each context. One of the best methods of checking the grammar and readability of your work is to read it out loud, as if to an audience. You might feel stupid doing it, but you can always shut yourself away so no-one will know! Reading out loud is also a good test to make sure that the structure as a whole makes sense. As you read, make sure that there are breaks in the text, and that they come in the right places for the sense of the text.

3. Get someone else to read it

Perhaps the easiest way to see if what you have written makes sense is to see if somebody else can understand it. It helps if it is someone who has a similar level of understanding of the subject as you, but a friend or family member will do. It is highly unlikely to be considered collusion if you swap assignments with someone else on the course solely for the purpose of proofreading each other's work. It is much easier to review other people's work critically, and you may well learn something about what does (or does not) work when communicating the topic. Just make sure that you do not go beyond this and start copying each other's work, which is likely to be flagged as academic misconduct (collusion—see Chapter 9, section 9.5.2).

A guide to punctuation

Punctuation often causes students significant problems and written work may end up with commas, semi-colons, and other symbols scattered around seemingly at random. As we saw with the case of the unfortunate bartender being shot by the panda, however, the presence or absence of a well-placed comma can completely change the sense of a sentence.

The full stop

The full stop is probably the easiest form of punctuation to deal with. Full stops are used to indicate the end of a sentence and are followed by a capital letter, which marks the start of the first word of the next sentence. Full stops are also traditionally used to indicate the use of a common abbreviation such as 'etc.' for 'etcetera', 'i.e.' for 'id est' (meaning, that is), and 'e.g.' for 'exempli gratia' (literally meaning, free example). In these cases, the full stop is not automatically followed by a capital letter. This usage of full stops for common abbreviations is decreasing and many writers simply put 'eg'.

The colon

The colon is most commonly used in the following ways:

- as in the previous phrase, to introduce a list or an example: 'A typical use of the colon would be as follows: …';

- to introduce a quotation: 'In his lecture, the professor stated that: "the nervous system …" ';

- to link two contrasting statements that might otherwise be written as separate sentences: 'The peripheral nervous system conveys information rapidly: this contrasts with the endocrine system …'. Here these two statements could have been written as separate sentences, but the use of the colon emphasises the contrast between them.

- to provide justification for a statement: 'The peripheral nervous system conveys information rapidly: this is especially important when the body is responding to damaging stimuli'.

The comma

The comma is the most widespread form of punctuation and is frequently misused. Commas are used primarily as a means of providing a pause in the flow of the text; the pause can be in a variety of contexts, which are probably best illustrated by examples.

Separating items in a list

Commas can be used to separate items in a list: 'Oestrogen, progesterone, follicle-stimulating hormone, and oxytocin are all hormones that …'.

Separating out distinct phrases

Commas are used to separate out distinct phrases or clauses within a sentence. For example, we might write the sentence: 'Testosterone, which is one of the sex hormones, plays a role in the development of lean muscle mass.' In this context, the sentence would still be correct and have meaning if the clause 'which is one of the sex hormones' was not there. However, its inclusion adds further qualification to the statement.

The presence or absence of commas in the sentence can also change the meaning of that sentence. For example, compare the following two sentences:

> The adrenal hormones which are associated with the control of blood pressure, act on both the heart and blood vessels.
>
> The adrenal hormones, which are associated with the control of blood pressure, act on both the heart and blood vessels.

These two sentences have subtle, but important differences in meaning. In the first sentence, the implication is that, of the hormones released by the adrenal gland, only those that are associated with the control of blood pressure act on both the heart and blood vessels. By contrast, the second sentence implies that all the adrenal hormones are associated with the control of blood pressure, and act on both the heart and blood vessels.

Creating pauses

Commas are used to create a pause after the opening phrase of a sentence: 'In this essay, comparisons will be drawn between ...' or 'As was stated earlier, the endocrine system is involved ...'.

This is by no means an exhaustive list of the usage of the comma and we would encourage you to follow this up by reference to the more detailed expositions found in texts on English grammar, some examples of which are given in *Further reading* below. As we have seen, the use of commas is important in imposing the correct structure and meaning to the sentences we write. To check whether you have placed your commas correctly, try reading aloud what you have written and ask yourself if the pauses indicated by the commas feel as though they are in the right place. This is by no means foolproof, but it will give you a useful guide.

The semicolon

The final common form of punctuation is the semicolon. Semicolons are employed much less frequently than commas. The semicolon can be thought of as being halfway between a comma and a full stop. It is typically used in the following ways:

To separate items in a list where the sentence structure is complicated and commas are already in use to separate qualifying phrases, for example:

> The hormones that regulate blood pressure include: adrenalin, which is released by the adrenal glands and increases heart rate and peripheral resistance; angiotensin, which is released by the kidneys and ...

To link two related sentences together, rather than separate them with a full stop, for example:

> Adrenalin is released by the adrenal glands; it has to be transported in the bloodstream to the heart.

The apostrophe

The apostrophe needs special mention in any consideration of punctuation because it is probably the most abused of any piece of punctuation, although it is actually fairly straightforward to use. Increasingly, the apostrophe can be seen randomly used before or after the letter 's' when the 's' is simply indicating a plural. This is incorrect. Apostrophes have two main functions: to indicate possession and to indicate contraction.

Possession

To indicate possession, the apostrophe is added to the word followed by the letter 's', thus: 'The heart's blood supply is …'. The test here is whether the sentence could be rephrased using 'of': 'The blood supply of the heart …'. If the word is a plural, already ending in 's', for example 'The kidneys', then the apostrophe is placed after the 's': 'The kidneys' blood supply is …' (do the test: 'The blood supply of the kidneys is …'). If the word is singular, but already ends in an 's', as is seen with some names such as 'James' then you still obey the rule for the singular form and add both an apostrophe and the 's', thus: 'James's heart condition worsened …'.

Contraction

An apostrophe can be used to indicate a contraction where some letters have been omitted because two words have been joined together to make a shortened form, such as 'don't' for 'do not' (remembering, of course, the point that you should normally avoid such contractions in academic writing!) There are, however, some traps for the unwary:

it's who's

These both indicate contractions: 'it is' and 'who is', e.g. 'It's clear that the experiment …'

Its whose

These are both possessive pronouns: 'The body regulates its blood pressure …'. Other possessive pronouns that don't take the apostrophe are: his, hers, ours, and yours.

Try this—Try out your punctuation

Now that you have read through these brief sections on punctuation, try using the rules to punctuate the following section of text. (Before you criticise us for not including any references in this piece, the references have been deliberately left out to simplify the exercise).

The peripheral nervous system comprises neurons whose cell bodies are located in the ventral horn of the spinal cord for the motor neurons and the dorsal root ganglia for the sensory neurons the axons of the motor neurons leave the cord via the ventral root and join the axons of the sensory neurons to form a mixed nerve these nerves branch along their route from the spinal cord to the tissues and end by innervating the peripheral structures such as the muscles and skin some sensory axons terminate as free endings within the muscles and skin whereas others innervate specialised sensory end organs in the skin these organs include the Merkel disks Pacinian corpuscles and Ruffini endings in the skeletal muscles the large diameter sensory axons innervate specialised proprioceptors called muscle spindles and tendon organs the axons of the motor neurons innervate the muscle fibres of the skeletal muscles with each motor axon innervating several muscle fibres but each muscle fibre only being innervated by a single

(Continued)

motor axon the axons terminate in a specialised structure termed the neuromuscular junction action potentials are conducted along the axon until they reach the pre-synaptic end of the axon here the action potential causes the release of the neurotransmitter acetylcholine (ACh) that diffuses across the synaptic cleft and binds with receptors on the post-synaptic membrane the binding of the neurotransmitter causes the muscle fibre to depolarise and fire its own action potential which results in muscle contraction.

Abbreviations

Abbreviations are often used in scientific writing, usually to remove the effort from writing out lengthy terms in full each time they are used. The usual rule of thumb is that you should write out the term in full, followed by the abbreviation in brackets, the first time you use it; on subsequent occasions you can just use the abbreviation:

> The electrocardiogram (ECG) was recorded while the patient .... The ECG recordings showed that ...

In scientific convention, some abbreviations such as 'ATP' or 'DNA' are so commonplace that they are often used without previously being written out in full. However, the safest practice is always to write the term out in full on first mention.

Further reading

Seely, J. 2007. *The Oxford A–Z of Grammar and Punctuation*. Oxford: Oxford University Press.
Truss, L. 2007. *Eats, Shoots and Leaves: The Zero Tolerance Approach to Punctuation*. London: Profile.

> Writing style exercise: suggested structure
>
> The peripheral nervous system comprises neurons whose cell bodies are located in the ventral horn of the spinal cord for the motor neurons, and the dorsal root ganglia for the sensory neurons. The axons of the motor neurons leave the cord via the ventral root and join the axons of the sensory neurons to form a mixed nerve. These nerves branch along their route from the spinal cord to the tissues and end by innervating the peripheral structures, such as the muscles and skin. Some sensory axons terminate as free endings within the muscles and skin, whereas others innervate specialised sensory end organs in the skin; these organs include the Merkel disks, Pacinian corpuscles, and Ruffini endings. In the skeletal muscles, the large-diameter sensory axons innervate specialised proprioceptors, called muscle spindles and tendon organs. The axons of the motor neurons innervate the muscle fibres of the skeletal muscles, with each motor axon innervating several muscle fibres, but each muscle fibre only being innervated by a single motor axon. The axons terminate in a specialised structure, termed the neuromuscular junction. Action potentials are conducted along the axon until they reach the pre-synaptic end of the axon; here, the action potential causes the release of the neurotransmitter acetylcholine (ACh) that diffuses across the synaptic cleft and binds with receptors on the postsynaptic membrane. The binding of the neurotransmitter causes the muscle fibre to depolarise and fire its own action potential, which results in muscle contraction.

11.2.3 Academic expectations (e.g. referencing)

There are some rules to scientific writing that are often not explained clearly to students. Markers will be able to spot them immediately, but it is often not clear to themselves where these expectations have come from. We covered some of these rules in Chapter 8, section 8.1.1, regarding academic voice and style. The main one that you should be aware of, and take particular care over, is appropriate referencing. This is often part of the marking criteria and something that some lecturers take very seriously. This section deals with some minor points on formatting for scientific writing and provides a detailed guide on formatting your references.

Figures (and Tables)

You should always think about using figures in your coursework, especially for visual and oral assignment formats like posters and presentations. Figures can often be used to help you explain a process or describe a structure much more effectively than by using words only.

You may be better off creating your own diagram, as it will be more specific to the point you are trying to make. For example, you could make:

- flow charts to indicate the operation of a control system as shown in Figure 11.1;

- diagrams to help explain a life cycle or a biochemical process;

Figure 11.1 Flow chart illustrating the feedback control of the release of growth hormone by the pituitary

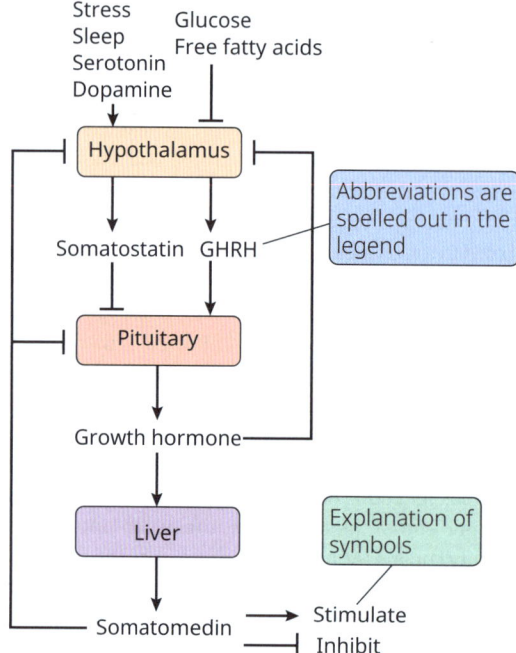

- drawings to help describe anatomical structures or the features of an organism;
- drawings of apparatus described in practical reports.

There are some key points to remember when including figures (and tables).

- The rules of plagiarism apply as much to figures as to the written word. If you copy a figure/table from somewhere to use in an assignment make sure that (a) you are allowed to do so [it should have a Creative Commons licence that allows its reuse with attribution], and (b) you cite where it has come from appropriately.
- Give each figure/table a number and refer to it in the text, don't just leave it as an 'add on'
- Give each figure/table a title so that the reader can see what is being shown.
- Make sure that your drawings are labelled so that the reader can identify the different structures being shown.
- Feel free to use digital software or freehand for producing your drawings: don't forget that in examinations you will probably have to draw your diagrams by hand.

Remember also that illustrations can save you words, so if you are struggling to write your essay within the word count, then a well-chosen illustration can save a lot of words in description or explanation.

Citations and references

When you are writing or presenting any piece of work, it is very important that you acknowledge the sources of the information and ideas that you are presenting. If you present an idea or a piece of information without acknowledging a specific source, then you are effectively claiming ownership of that idea or information. Provided it is your idea then that's okay: for example, it could be your interpretation of the results obtained in a practical class or from your research project. The failure to acknowledge sources, however, is one of the most common reasons for being accused of plagiarism (covered in detail in Chapter 9).

In this section, we will look at a standard way of referencing, and discuss when and how you should do it.

Defining the terms

The process of acknowledging sources sometimes seems more complicated than it need be because of the different terms used to describe what you should do. We will work through some examples, but it is useful to explain some of the terms first.

Referencing is the name given to the process of acknowledging any material in the form of ideas, specific information, drawings or illustrations, experimental methods, computer programs, etc., that has come from someone else's work. Normally, this is work that has been published in some form, for example as a research paper, book, or film, but it may also be simply a theory that someone has told you.

Citations are the specific acknowledgements to other people made, for example, in the text of your writing. Normally, these will be the name of the person and the date when their material was published.

References are the list of sources that you actually referred to in your work. Every citation in the text should appear in the reference list and every source listed in the references should refer to a citation in the text.

Bibliographies are lists of all the sources that you made use of during your research. A bibliography may, therefore, include sources such as general textbooks that you used for background reading, but from which you did not actually take specific material for your work. Some of the material listed in the bibliography may not, therefore, appear as a citation in the text.

For most work at university level in the Biosciences, particularly in the later years of your course, you will be expected to produce a reference list, rather than a bibliography.

Why should I cite other people's work?

The main reason, as indicated in the introduction to this section, is so that someone reading your work will know where the ideas you are presenting came from and will be able to identify which ideas or pieces of information that you are presenting are specifically your own. The references are also very helpful to the reader who needs to know more about the background to the topic. He or she can go to the articles you have cited to find out more detailed information. When you are preparing coursework, the list of references will show your tutor what materials you have used, and he or she may be able to give you feedback about the suitability of the materials and your search strategies.

Do I have to cite everything?

The general principle is that anything that you include in your work that is taken from someone else must be acknowledged. This is true just as much if you are quoting someone word for word, or if you are taking their ideas and putting them in your own words. Having said that, there is a level of information that is now accepted knowledge and for which it is not necessary to provide a citation. For example, if you write the statements:

> Chlorophyll is the main pigment involved in photosynthesis in green plants.
> Bats are mammals that can fly.
> Haemoglobin is the respiratory pigment involved in oxygen transport in humans.

These are accepted facts and there is no need to cite the source of the information (although you could cite the textbook from which you derived the information). Indeed, the history and origin of such information is often unclear anyway! To get a feel for how much you should cite, one of the best things you can do is think about it when you are doing your background reading—when you read a review article or a research paper, notice the level of information that the authors are taking as accepted and the levels at which they make specific citations. Have a look at these guidelines as a framework for knowing when to cite your sources.

You should always cite the source when:

- you are quoting word-for-word from a source, whether it is a textbook, research paper, lecture, or film. Any such quotation must also be written in quotation marks;

- you are copying, or taking elements from, a figure from a piece of published work;
- you are presenting information or ideas that are critical to the arguments you are making;
- you are presenting any ideas that may be controversial, for example, someone's theory about a specific process;
- you are presenting information based on specific experimental or observational evidence.

If in doubt, then acknowledge the source.

How to draft citations and reference lists

If you look through different academic journals, and particularly if you look at publications in different subject areas, you will see that the citations and reference lists are formulated in a variety of ways. There are several standard styles for referencing such as Harvard, Vancouver, Chicago, and so on. Although the styles are different for these, the basic information is always the same. In this context, Harvard is one of the most common formats for citations and reference lists in the biosciences, so this is the format we will employ in this text.

In the Harvard system, the citation in the text simply comprises the name(s) of the author(s) and the date. In the reference list, you must also include the author's (or authors') initial(s), the title of the article and the journal or book it comes from, the volume number, and the page number. In the following examples, we will run through the most common variations for citations and references. Note that, even within the Harvard system, there are differences in presentation. The following formats are based on the British Standard (BS ISO 690:210), but you should check with your lecturers whether your department recommends any specific styles. In the sections below, we cover the main principles of using the Harvard system. Your university library or learning resource centre should have guidance on referencing techniques.

Referencing journal articles

For much of your study, articles published in scientific journals will represent the major source of information. Note that the following formats for referencing apply to all journal articles, even if you downloaded the article from the journal website.

Citation:

If there are one or two authors for a given publication, then all the authors' names are given:

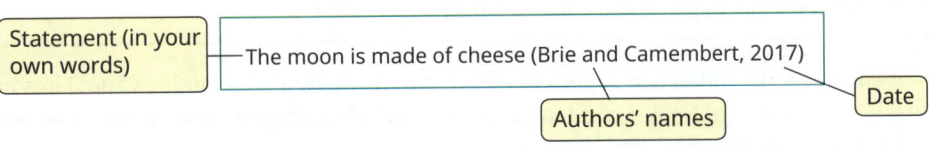

Brie and Camembert (2017) reported that the moon is made of cheese.

Reference:

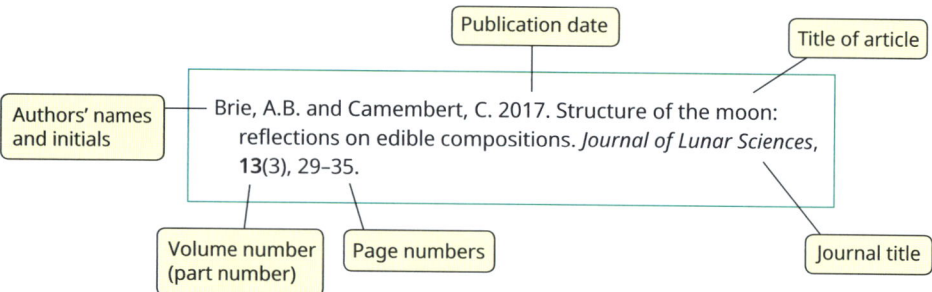

Brie, A.B. and Camembert, C. 2017. Structure of the moon: reflections on edible compositions. *Journal of Lunar Sciences,* **13**(3), 29–35.

Labels: Publication date · Title of article · Authors' names and initials · Volume number (part number) · Page numbers · Journal title

If there are four or more authors, only the first name is given in the citation followed by 'et al.' (this comes from the Latin expression *et alii*, meaning and others), but all the authors should be listed in the reference. If there are three authors, then all authors should be included in the first citation, and thereafter the format is the same as for four authors.

Citation:

> The moon is made of cheese (Brie et al., 2017).

OR

> Brie et al. (2017) reported that the moon is made of cheese.

When giving a direct statement some authors prefer to use a phrase such as:

> Brie and co-workers (2017) reported that ...

To cite two different papers by different authors in relation to the same statement:

> The moon is made of cheese (Brie and Camembert, 2017; Cheddar, 2018).

OR

> Brie and Camembert (2017) and Cheddar (2018) reported that ...

If you have more than two papers by different authors, listing the authors and dates in brackets is usually preferable to a long list of names in the body of the text. Note that the publications are normally listed in chronological order.

To cite two papers by the same author(s) in different years, list the papers in chronological order:

> The moon is made of cheese (Brie and Camembert, 2015; 2017).

To cite two papers by the same author(s) in the same year, the papers should be identified as 'a' or 'b' and this distinction added to the reference list.

Citation:

> The moon is made of cheese (Brie and Camembert, 2017a,b).

References:

> Brie, A.B. and Camembert, C. 2017a. Structure of the moon: reflections on edible compositions. Journal of Lunar Sciences, 13(3), 29–35.
>
> Brie, A.B. and Camembert, C. 2017b. Variations in the density of cheese on the lunar surface. Journal of Lunar Sciences, 14(1), 54–63.

The assumption so far is that these journal articles are articles that you have read, i.e. that these are primary sources. It may be that you wish to cite the work of a researcher but have not read the material in the original article—for example, if you have not been able to obtain a copy of the article. In this case, you may have found that the information you need is quoted in a more recent paper by another author, and so you can refer to this more recent paper. This is clearly more indirect than taking your information directly from the original research paper, and it is referred to as a secondary source. The use of such secondary sources is not ideal, but sometimes it is unavoidable, particularly when you cannot access the original article. The format for such referencing is as follows (note that the reference is for the source from which you extracted the information, not the original source).

Citation:

> Cheddar had originally proposed that the composition of the moon was of blue stilton (Cheddar, 1994, as cited in Brie and Camembert, 2017).

Reference:

> Brie, A.B. and Camembert, C. 2017. Structure of the moon: reflections on edible compositions. Journal of Lunar Sciences, 13(3), 29–35.

Referencing books

If you wish to cite material from a book, the citation in the text takes the same formats as for the journal articles: the name(s) of the author(s) and the date of publication. The reference list format varies according to the type of book. The basic format is as follows:

> Brie, A.B. and Camembert, C. 2018. *The Composition of the Moon*. London: Lunar University Press.

Many books are compositions of articles that have been edited by the main editors. In this case, the reference would be:

> Brie, A.B. and Camembert, C. (eds)   2018 *The Composition of the Moon*. London: Lunar University Press.

In the case of such edited books, you may have only cited a specific chapter in the book that was written by another author. The format for such a reference is as follows:

> Edam, G 2018. Studies of the lunar surface. In: Brie, A.B. and Camembert, C. 2018. *The Composition of the Moon*. London: Lunar University Press, pp 25-36.

Referencing websites

An increasing amount of information is available from websites and many such sites can be very valuable sources. As with any other source of information, that source has to be acknowledged through a citation in the text and a reference in the reference list. The citation in the text is, as for the other types of source, author(s) name(s) and date of publication. The basic format for the reference list is as follows:

> Lunar Development Agency. 2018. *Mining the Lunar Surface* [online]. Available from: http://www.lunardevelopemnt.co.uk/htm [Accessed 01 April 2018].

Digital object identifier (DOI)

The DOI is a means of identifying specific links to articles on the net. When an article is published and made available electronically, the publisher will give it a DOI number that can then be used to locate the article. When you draft your reference list, it is good practice to include the DOI for print-version articles, as well as those that are only published on the web. In this section, we have looked at the styles of citation and referencing for the most common sources of information that you might be using for your work. As stated before, there are different formats for presenting such information and, here, we have concentrated on the Harvard format, which is one of the standard forms used in the Biosciences. You should, however, ensure that you check with your university or college to see if they have a preferred style of referencing.

Final checks before submission of the assignment with respect to referencing

1. Make sure that your key statements are supported by citations in the text;
2. Check that each citation in the text is matched against a reference in the listing at the end of the essay;
3. Likewise, check that each reference at the end appears as a citation in the text.

 ## Chapter summary

This chapter has focused on the process of iteratively creating and improving a draft assignment before it is submitted. It is important that you leave an appropriate amount of time for this process if you wish to maximise the quality of the work you submit for your degree. The chapter provides specific questions that you can attempt to answer as you review your submission as a whole and on a line-by-line basis. We have also provided specific guidance on expectations of how work is written, formatted, and presented. These include professional expectations (such as the use of good grammar) and academic expectations (such as correctly referencing the research that has informed your content).

This marks the end of a long journey, the process of preparing for, planning, researching, drafting, redrafting, and submitting a single coursework assignment for your university degree. It is entirely possible that you will have to repeat this process upwards of 20 times throughout a three-year course. It is a challenging process, but it is where you get to really develop and evidence the skills and competencies the degree is designed to develop. With respect to this book, we recommend that you return to Chapter 7 at the beginning of the creation process for each coursework assignment you have on your course. It is also important to note that submission of an assignment does not mean that you will never have to think about it again. In a few weeks, you will probably receive some important information: the grade the work has achieved, and feedback on what you did well and where you need to improve on future assignments. Reflecting on this feedback as you complete future assignments is as an essential element of your degree as attending lectures. When you receive the feedback on the assignment you are working on, have a look at Chapter 13 for guidance on how to make the most of it. It can be helpful to view this as a cycle of progression, each time you should reflect on the progress you have made (which will be helpful for Chapter 14 *Making yourself employable*).

Chapter 12 is the final chapter on assessment in this book and deals with two skills that are critical to university success but do not apply to every assignment: collaboration and critical thinking.

12 Key competencies: collaboration and critical thinking

→ Introduction

Two of the most important competencies you will develop at university (on any degree) are collaboration and critical thinking. Both skills are highly transferable to your personal and professional lives and are commonly at the top of lists of things employers want to see in graduate applicants. In fact, collaboration and critical thinking are two of the four skills students 'must master to succeed in work and in life' according to the Partnership for 21st Century Skills (the other two are communication and creativity, which are covered throughout the rest of the book).

Despite the importance of both collaboration and critical thinking skills, they have historically not been taught directly by lecturers and many people expect them to develop naturally over the process of the degree. Collaboration will commonly contribute to assessment on your course directly through group assignments or indirectly, for example, when working together during practical classes. Critical thinking becomes an increasingly important component of assessment marking criteria as you progress through your degree. This chapter has been designed to give you some foundational tips, techniques, and processes to help you actively develop, apply, and evidence collaborative and critical thinking skills in your university assignments (and beyond).

12.1 Group work: making it work for you

The process of collaborative, or group working, may often be very productive. Indeed, there is a wealth of evidence that material learned through group work is retained better than that learned through formal teaching methods. Group work is often very useful in developing critical skills because, as a member of a group, you need to evaluate the contributions that your fellow students have made, as well as reflecting on your own contributions and discussing their validity with the rest of the group. As an example, this book was written as a collaboration between multiple authors and editors, who worked together to ensure that the content made sense to as many people as possible and pooled ideas from people with different perspectives and ideas.

In the Biosciences, group work is very common as a part of practical classes where you typically may have to work in pairs or threes in order to complete the experiment. It is very important to be clear about the extent of the collaboration, e.g. in a practical class you may be required to work together to obtain the data, but then to write individual reports based on that data. In another exercise, you may be asked to work together to produce a written report, but also draft out individual reflective commentaries on the process. In these types of exercise, be very clear where the boundaries lie between the collaborative and the individual elements, so that you don't fall foul of your university's regulations on collusion (Chapter 9, section 9.5.2 *Collaboration and collusion*). For the purposes of this chapter, we will assume that you are being given fully collaborative exercises to undertake and that there will be a single product, produced by your group, as the outcome. Note, however, that it is possible that marks will be allocated individually using a peer-review process. The lecturer setting the work should explain how this works and what effect it will have on the final grade (if any).

The organisation and dynamics of interaction of your group will be critical to its success and will be strongly dependent on everyone being willing and able to support the group as a whole. Assessments requiring group work exercises will normally require you to work with your group on a set task as an extended exercise that you undertake at times and locations of the group's choosing. In some exercises, you may be required to work online without any physical meetings: this may be through an online conferencing tool such as Microsoft Teams or Zoom, where you can see each other, or it may be through text-based interactions where you have to follow threads. However you are engaging with the group, the dynamics of how people interact and contribute to the group are a very important element contributing to the overall success of the group. Under these conditions, how you behave in the group is also clearly important. Some key points to remember here are:

- keep an open mind and be prepared to listen to, and consider, the ideas put forward by others in the group;
- avoid being judgemental—for example, by being dismissive of someone else's ideas;
- be careful of your body language—for example, folding your arms and staring out of the window does not convey a positive message;
- if you are engaging virtually through text alone, take care how you phrase your statements—when you are trying to express views succinctly through text, it is often also very easy to come across as being brusque.

Let's now consider each of these types of group activity.

Short-term group work

Short classroom exercises are often employed as a way of breaking up lectures or as a means of checking that students have understood a specific concept. You will probably be asked to talk to the people around you to discuss a specific topic or answer some questions. Many of the issues of group dynamics and teamwork that are discussed in the next section are not applicable here. For the most part, the most important skill is remaining focused on the topic and using the time constructively—it is very easy for discussions to degenerate into a review of the latest gossip and, although that might be of great interest, it will not help get the task completed! It is also often very easy to spend a long time debating the first one or two aspects of the question and then find you have run out of time before completing the exercise, particularly if you only have a few minutes for your discussion.

There are several ways of providing some structure to your discussions and making sure that you get some answers, while also getting everyone involved as far as possible.

If you have been given several questions to answer, it is very helpful to agree quickly at the start how long you are going to spend on each question, and make sure that you adhere to that timescale, noting down your agreed answers as you go.

A useful way of getting ideas quickly and of engaging everyone in the discussion is to go round the group in turn with everyone putting in an idea or point of information related to the question.

An alternative is to generate ideas. People in the group mention ideas in any order as they think of them. This can be very productive if some creative thinking is needed, because the lack of formality encourages more open thinking. It can also generate numerous ideas very quickly. However, it may also lead to some people feeling excluded because, for example, they feel apprehensive about the merit of their ideas or because they are too shy to contribute. In such situations, try to be aware of the other members of the group and encourage non-contributors to participate: after all, they might have the best ideas of everyone! As with the guidance for behaving in tutorials, the key points here are to listen to what others have to say, to ensure your body language is appropriate and open, and not to be dismissive, even if you disagree with what the other person is saying.

Longer-term group work

If you have been allocated a task that will require you to work in a group over a period of days or weeks in order to complete the task, then the dynamics of the group become a very important factor contributing to the success, or otherwise, of the assessment. Many students are dubious about the benefits of group working, particularly if the assessment is worth a significant number of marks. However, when you get to the stage of applying for jobs and being interviewed you will doubtless be asked questions about your ability to work in a team, what kind of role you play, and to give examples of when you've had to work hard in a team context to complete a shared task. Pointing to experiences of sports teams, or your involvement in clubs and societies, or part-time work are all good examples, but being able to give examples from an academic context is much more relevant to most professional applications.

Undertaking a specific project as a group can be broken down into a series of key stages, as shown in Figure 12.1. We will consider each stage in turn.

12.1.1 **Get off to a good start**

It may be that you have been allowed to select who you work with and have chosen to work with a group of friends. On the other hand, you may have been put into groups by your lecturer, and you may or may not know the people you are having to work with. You might not feel happy if you find yourself grouped with people you don't know, or perhaps don't like, but it can be very good experience for your working life, since you may well have to team up with strangers to undertake a task in your employment, or you may have to work with people you don't necessarily like. In either case, the work still must be done. Time spent in ensuring that the group gets off to a good start should pay dividends as the project develops.

Figure 12.1 Stages in the process of group work for a specific project

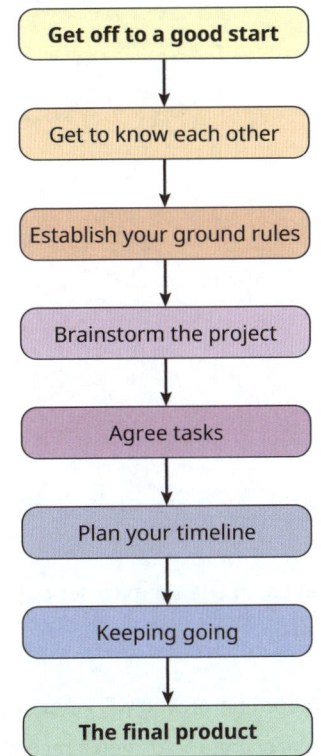

12.1.2 **Get to know each other**

This first stage is clearly most important if you are going to be working with people you don't know. Even in a group where you know everyone, that might not be the case for everyone else. So, it is very helpful to get together for your first meeting, away from the work environment, perhaps in a coffee shop, or just sitting on the grass if the weather is good. Use that meeting to ensure that everyone feels part of the group and can, at least, begin to feel confident about working together. If you need to do a round of introductions, get each person to say something about themselves as a 30-second biography or say what they think might be particularly interesting about the project.

12.1.3 **Establish your ground rules**

Don't be misled into thinking that, just because you are working with friends, you can all just get on with the task. Whether you are with friends or strangers, it is essential that you agree the ground rules for running the project, otherwise things can come unstuck very quickly. Working collaboratively in a professional environment doesn't just 'happen', there are roles and processes that help group-based projects run smoothly (which can be helpful to evidence in your CV). The following questions/points, provide a useful starting point for

Figure 12.2 Proposed plan for the group's first meeting

BioStars Project Group Meeting Plan
1430 Tuesday 2 May
Biology dept coffee bar

1. Introductions
2. Group organisation
     a) Do we need a group leader?
     b) Do we need a secretary?
     c) Do we need . . .
3. Group operation
4. Brainstorm
5. Division of project tasks
6. Key progress dates
7. Next meeting

professional group working, you could use them to set out a generic agenda for the first meeting (as shown in Figure 12.2):

Group roles:

- Do we need a project leader/coordinator who will keep track of progress?
- Would someone take the role of secretary to keep a record of points agreed at meetings (as shown in Figure 12.3), and make sure that meetings stick to time? (It doesn't have to be the same person each time.)
- What time commitment should each member of the team be making to the project?
- What action should be taken by the group if one or more members are not contributing?

Collaborative process:

- What technology shall we use to store our joint work? Local files on individual computers are not collaborative and may make the group over-reliant on one individual. Consider using more collaborative software like Microsoft SharePoint or Google Drive, where 'live' files can be accessed and edited by the group members (other examples include Dropbox or your university's virtual learning environment).
- What channels will we use to communicate? University email is helpful as everyone will have one and it doesn't require exchanging personal details. Having everyone agree a channel at the beginning can help address issues later on where a group member feels left out or says they have not been receiving communication.
- Do we need formal agendas for future meetings (the answer is probably yes)? If you think that setting agendas makes the whole process seem too formal, then you could have meeting plans. At the very least, there should be a list of agreed actions and who is supposed to complete them at the end of each meeting!

Figure 12.3 Example of first section of notes from the group's first meeting

BioStars Project Group 1st Meeting Report
1430 Tuesday 2 May
Biology dept coffee bar

1. Present
Alex, Komal, Sarah, Lee, Rathan

2. Organisation
Komal will act as group leader, Sarah will take notes

3. Operation
Project task is to present a critical review of deep brain stimulation in the treatment of
Parkinson's disease. 6 weeks working time.
Group to meet every week to update

4. Brainstorm
PD - degeneration of dopaminergic neurons in substantia nigra (basal ganglia)
movement deficits . . .

12.1.4 Generating ideas

You should have been given the project title and a brief for what is expected of you (make sure it has been analysed thoroughly (Chapter 7). No doubt, different members of the group will have different ideas about what the approach should be, what needs to be done, and how. Use a free, unstructured session to let everyone have their say. It is important that the discussions at this stage are free ranging, so that the full breadth of the topic can be covered and also so that everyone feels free to participate. At this stage, all contributions should be treated equally and there should not be any criticism of the ideas that are offered: you will have time later on to focus on specific aspects. Note: it is important that someone records the ideas, otherwise all that brain power will be going to waste! However, you need to make sure that the person recording the ideas has the chance to contribute ideas as well and is not just there as a note-taker. Also, set a time limit for the discussion, otherwise you could be there a long time. A useful approach to generating ideas is to give everyone two or three minutes to think about the problem and then go round the room so each person puts their ideas in. These ideas can be recorded as bullet points on a flip chart or similar, which everyone can see. When the first round of ideas is up, see if there are links between them so that you can group them together through a collaborative critical reflection. From that grouping, you can then go round the room again to build on these with further ideas but remember your time limit.

Figure 12.4 Identification of the tasks involved in preparing the presentation

Tasks for Parkinson's disease project

Research topics

1. Structure & function of the basal ganglia
2. Neurological features of Parkinson's disease (PD)
3. Current drug treatments
4. What is deep brain stimulation? (DBS)
5. Use of DBS in treating PD

Identify the key points for the presentation
Prepare the presentation and write the text
Prepare handouts for the audience
Rehearse and review the presentation

12.1.5 Agree tasks

Once you have the outline of how to approach the topic, you need to break this down into a series of tasks, so that you can plan the way the work will be done and who will do it (as shown in Figure 12.4).

One of the obvious features of group projects is the division of labour so that there is co-operation; if you have everyone trying to do everything, there is no real benefit from being in a group. On the other hand, you need to make sure that everything is covered and that everyone has knowledge of the project as a whole. So, it is vital that, right from the outset, the group agrees who is going to do what and that a common record is kept of the division of labour (Figure 12.5). So, having drawn up a list of tasks from your ideas session, the group needs to agree how they are divided up so as to be reasonably equitable. At this stage, it is really helpful to take account of different people's particular skills—if you have a successor to David Attenborough as a presenter, then it makes sense for that person to be taking a major part in any presentations (although make sure that everyone presents something). Just make sure you understand how the work is graded and that people will receive appropriate reward for their efforts.

12.1.6 Plan your timeline

As well as dividing the project up into different elements that individuals will be responsible for, you also need to divide it up into chunks of time so that you have specific milestones that you can identify, in order to record progress. It may be helpful to link this in with the agreed division of labour, so that each person has a target to work towards and take responsibility for (Figure 12.5). Also, build in some slack in the timing, so that you allow some extra time at the end of the project for completion. For example, you could aim to complete the project a week before the deadline, then if there are any problems you do have some leeway.

Figure 12.5 Key dates for the project and breakdown of project tasks

Project timeline:

May 9th	Completed background research
May 16th	Agree the bullet points for the presentation
June 6th	Draft presentation complete
June 10th	Rehearse presentation
June 13th	Class presentation!!!

Work division:

May 9th

  Komal - neurological aspects of PD
  Alex - current drug treatments
  Rathan - what is deep brain stimulation?
      . . .

Try this—Practise planning

Think of a simple task with several steps that you are going to carry out by yourself, such as preparing a meal for yourself and three friends. Decide what time you want to eat and write down each step required with the time it needs to be done by. Repeat the process, but now assume that your friends are helping you to prepare the meal. Write down each step involved, the time it has to be completed by and by whom.

Keeping going

It is often the case that everyone will be pretty enthusiastic about the project to begin with; the first meeting goes well with agreed plans of action and so on. Keeping the momentum going, however, can be much harder. This is especially true when more imminent course-work deadlines focus input elsewhere and the long-term nature of the group project tends to give rise to an element of complacency. This is when it becomes important to monitor—and stick to—your original project plan.

Meet regularly

Keep to your schedule of regular meetings. Don't let a meeting be dropped because people have not done their work (even if one of those people is you!); it is very easy to let this happen, and you will quickly find you are on a slippery slope and no progress is being made. If nothing else, therefore, the meetings will act as a reminder that the work needs to be done and that the group needs to keep on schedule, otherwise the project will be compromised. This also adds incentive because people will be embarrassed to admit to the group that they have not done their work.

Keep a check on progress

At each meeting, it is important to review exactly where the project is and whether you are on track. Refer back to your timeline (Figure 12.5) and check off your progress against the target dates. This checking, however, is not just a tick-box exercise, accepting that the work has been done without being sure that it actually has. If the group is aiming for high quality in its performance, then there also needs to be an evaluation of the quality of the work each person has produced. If your tutors gave you a marking scheme or equivalent, then the members of the group need to evaluate each section of the work that has been produced against the marking scheme. Although it may seem like it, this is not just a chore—it is important for your own learning that you have an understanding of the whole project, not just the sections you are responsible for. It is also a very valuable learning exercise in developing the skills of critical evaluation, reflection, and discussion.

People who fall behind

One of the biggest, and most common, problems with group work is dealing with group members who don't keep up with the work or fail to engage with the group. There are various strategies the group can adopt to try to resolve the situation.

- Talk to the person: this is the first and most vital approach. If the person comes to the meeting, then you can discuss the situation. If he or she has stopped attending the meetings, then you need to try to re-establish contact as soon as possible.

- Find out what the problem is: it may be that the person has been ill, or is struggling with something else and can catch up, particularly if given a bit of time or some help as a one-off. If there is no real excuse, then the group needs to refer back to the ground rules and make it clear that everyone needs to provide their agreed contributions, otherwise the whole project may fail. Check to see whether a group extension on the deadline is a feasible solution to the issue.

- Persistent non-cooperation: if one member of the group will not cooperate, even after you have discussed the position, or if that person refuses to respond, then you need to take more explicit action. Probably, the best option here is for you to contact the lecturer and explain the situation, and ask for assistance in resolving the problem, although this should be an option of last resort as it indicates that the group can no longer sort out its own problems.

12.1.7 **The final product**

The deadline is approaching, and you now need to bring everything together ready for submission. If you have let things slip, then the quality of the final product may be lacking. If, however, you have kept to your timeline, this phase of the project should allow you to produce a really polished piece of work.

- Bring everything together: one of the difficulties of group work is that everyone has their own style of writing or expressing themselves, but the final piece of work needs to appear as a consistent piece, not as three or four completely separate elements that have been lumped together. This may require you to edit the sections so that the approach or style of writing is the same and also so that there are sensible links between the separate sections. Editing by committee is often a lengthy and unsatisfactory process. In most cases, it is preferable for one person to be responsible for bringing the sections together and editing them to create a single 'voice'—think back to the initial process of agreeing the tasks so that you could play to the strengths of each member of the team. This stage can be sped up somewhat by working on a single output collaboratively, making the most of digital resources such as Google Drive or OneDrive.

- Check against the brief: before you submit the work, run through the instructions you were given at the outset again, and make sure that you have addressed all the criteria. It is important that, as far as possible, everyone agrees that they are happy with the final product and are prepared to sign it off. When it's all over don't forget to make use of the feedback (see Chapter 13 *Making the most of feedback*) and think about where any problems were and how you might do it better next time.

12.2 Critical thinking and writing

As discussed at the beginning of the chapter, critical thinking is considered by many to be one of the essential 21st Century skills students must develop (alongside collaboration, communication, and creativity). As you progress through your studies you will find that 'evidence of critical thinking' becomes an increasingly important determinant of the grades you achieve on your assignments. By your final year, most courses will expect you to be able to independently investigate a subject, consider the different perspectives within that field, critically evaluate the evidence that supports those perspectives, and synthesise a conclusion drawn from a review of a broad range of literature on the subject. In this section, we outline two related but distinct types of critical thinking that will help you evidence criticality in your assignments: critical evaluation of specific sources, and critical synthesis of content from across multiple sources. Do bear in mind that, in common understanding, we tend to use the terms 'criticise' or 'critical' in negative contexts, implying that something is wrong. In this context, one of the definitions given in the Oxford Dictionary is really useful: being critical involves 'the objective analysis and evaluation of an issue in order to form a judgement'.

12.2.1 Critical evaluation of evidence

Many people (including lecturers, researchers, employers, and policy makers) view critical thinking as the single most important skill developed by a university degree. Humans have access to more information than they have ever had, more than could possibly be consumed in a single lifetime. However, not all this information is equally representative of reality, i.e. some of it is 'fake news' (a situation likely to get worse with the rise of Generative AI). Even scientific journals should be viewed with scrutiny, with an increasing number of predatory journals prioritising profit or looking to further a specific agenda. It is, therefore, imperative that we become able to critically interrogate the media we consume by independently identifying the quality of the evidence that supports key points and researching what alternative viewpoints there are. This skill has equal value for you personally and professionally, as many roles will require you to make difficult decisions based on review of the available evidence. From a personal perspective, think of political debates, news articles in the media, or advertisements. All of these set out to justify a specific position based on evidence, but it is very important that you don't take this on trust; you need to look critically at the evidence, how it was collected and analysed, and whether the interpretations are justified. Do most dentists really recommend this toothpaste and what does 'most' mean? Is the job market actually improving or getting worse? Will this herbal remedy actually make you better?

12.2.2 Is this an appropriate source (for referencing in your assessment)?

From your first year of study, you will be expected to distinguish between a 'good' source of information (e.g. a peer-reviewed scientific paper) and a 'bad' one (e.g. a biased, poorly supported website). We recommend that from your first assignment you should be focusing on only using 'good' sources of information to construct your assignments. So, what kinds of things make a source trustworthy?

Question 1: Who created the source and are they credible?

Can you find some independent verification of expertise online? For example, who do they work for? Have they published a lot of papers or earned qualifications in relevant disciplines? Most importantly, do they have a vested interest in making people see a problem a specific way or people believing a specific answer is 'true'? Is it even possible to find out who the author(s) are?

   If the author is not reliable, you probably do not want to include the source of information in your research. Try to find other information sources.

Question 2: Is the source presenting information on a subject that may have changed since it was published?

What date was the information published? Easy to find with scientific journals and text-books, but difficult on websites or online sources.

   If it is possible that the information is out-of-date, try to find a more recent source of information. If you cannot, make reference to the potentially dated nature of the source in your assignment. For example, an article from 1950 is fine if you are trying to illustrate what people thought about that subject at that time, but it has probably been superseded by more recent research. Older references are also essential where you need to reference the first instance a finding has been reported, or you wish to report how understanding of the subject has changed over time.

Question 3: Has the source been reviewed by impartial experts?

Has there been some form of peer-review of the information source by someone other than the authors? Scientific papers are (generally) reviewed by other experts in the field before they are published. So, in most cases there has been at least some impartial review of the contents. Has anyone checked your source to see if it is in fact unbiased and trustworthy?

If the source has not been peer-reviewed, then you may want to find different and more appropriate sources of information. However, if you cannot find the information anywhere else you can cite the source of information if you refer to the reason it is potentially inappropriate (e.g. it's a website that is attempting to spin facts in a specific way).

Question 4: Does the source provide data or citations to evidence the key points?

Most scientific papers make conclusions based on data collected as part of an experiment; this is helpful because you can check whether you agree that the data support the points being made. At the very least, there should be a reference to another source that contains data that support each point being made (as you would find in a literature review).

This is a tricky one, but sources that rely solely on other sources to support key points (for example Wikipedia) may be misrepresenting the findings or conclusions of other people. It is often best to go to the source being cited instead, where you can scrutinise the evidence yourself. Wikipedia can be a helpful place to start reading up on a subject as it can direct you to some helpful sources on the subject, but Wikipedia should not be the final place you look for information.

Most of these questions are much easier to answer of scientific literature than other information sources, which is one of the reasons you are expected to focus on them in your research (see Chapter 9, section 9.3.1 for a list of 'appropriate sources' of information). However, one cannot always assume these things due to a rise in journals that will publish anything if the author can pay the publication fees. Also, note the existence of pre-print servers (such as bioRxiv.org), where researchers can put scientific reports online without peer-review process while they undergoing the long process of trying to get their work in a scientific journal.

If the source you have found passes these checks, then you can probably use it to support points being made in your assignment. It is likely that these criteria will exclude most websites or science communication articles, which are rarely appropriate sources of evidence to support key points in your assignment (best to go to the original source of information). A key caveat to this is if you want to include an untrustworthy article as an example of how an argument is poorly supported or aligned to biased interpretation of the subject, just make sure you are explicit about that in your assignment. If you are a first-year student, this is probably an appropriate level of critical evaluation, but students further on in their journey will need to show more criticality.

12.2.3 Extracting key points from a research paper

As you progress through your studies, you are going to be expected to read an increasing amount of scientific literature, both as guided reading and research for your assessment. It is important, then, that you can rapidly extract the most important information from a scientific paper as quickly as possible. Below, we provide some questions you should attempt to answer when reading a scientific paper and what parts of the paper you should look for that information in.

Key finding—what are the most important conclusions the authors have made in this study?

The study Abstract will probably include the most important finding of the paper. If not, you might be able to find this in the Conclusions or end of the discussion section.

   This will probably be the key point, fact, or statement that you cite this paper for in your assignment. If all you need to know is the key thing that the study found, then you can probably stop here. However, if you need to understand the paper in more detail (i.e. in order to critically evaluate it), then you should continue.

Question 1: What research questions has the study been designed to investigate?

The Abstract should have a sentence that attempts to explain what the authors were trying to answer. Also, check near the end of the Introduction section (it may even have a subheading called Research Questions, Study Aims, or similar).

   If you want to critically evaluate a study, you need to know what it was hoping to achieve.

Question 2: What does the paper suggest the answers to the research questions are?

Start with the Abstract that may provide this information. Then check the opening section of the discussion, which should attempt to place the study results in the context of the research questions or hypotheses. This may be the same as the Key Finding statement above.

   In order to effectively evaluate the paper, you need to be clear on exactly what the authors are saying their study has found. You can then investigate the rest of the paper to see if you agree.

Question 3: How does the paper suggest that our understanding of the subject has changed as a result of the answers they have found?

The Abstract should have this stated towards the end. There may also be a Conclusion section that should have this stated clearly. Reading the rest of the Discussion may provide additional clarity. Again, this may be the same as the Key Finding statement above.

   A common issue with scientific papers is a tendency to 'overstate' the results in order to increase the importance of the paper. You can consider whether the results of the study actually support what the authors say or not.

   If you can answer all four of these questions, you probably have a good enough understanding of the paper for inclusion in an assessment or discussion in a tutorial. However, critical thinking will require further steps, so carry on with the next sections for more advanced questions to find answers to. If there are any steps that you cannot write the answer to, make note of them. You could discuss these points with colleagues on the course, with your personal tutor, or with the module leader. The more you do this, the better you will get at it. It's very difficult, as you are unlikely to be the target audience for most papers, and they assume a large amount of experience on your part. Just try to do your best and evidence what you can.

12.2.4 Critically evaluating a research paper (is it good science?)

At some point in your second year, you are likely to be expected to critique even good sources of information, which means being able to effectively evaluate scientific papers. This may take the form of a specific assessment where you must critique a paper given to you. Mostly commonly, you will have to evidence that you evaluated the quality of evidence supporting the points in your coursework. Below, we outline some questions you should attempt to answer when reading the paper that will help you with your critical evaluation (note, you need to complete the questions asked in section 12.2.3 and then answer these questions).

Question 1: What experiment has been designed to answer the research questions?

The Abstract will contain a very brief summary of the kind of data collected. You will need to look at the Methods section if you need to consider this in detail.

You are looking to see whether the methods selected will produce data that has the potential to provide answers to the research question. If they do not, then you can point this out in your assignment.

Question 2: Are there any limitations of the method that mean the answers cannot be applied to a broad range of contexts?

This will require you to look at the Methods in detail and compare them to the key research question being asked (dealt with in one of the earlier questions). Focus on the section that explains the study system (which will most likely be at the beginning). Try to write down how the researchers have sampled the real world. Remember it's not possible to collect data from every relevant system for forever, so the study must have considered a subsample of what is available.

One way of evaluating the study is to consider what are the limitations of the study sample as a representation of the larger, more complex system. Some example limitations of sampling method include:

- The study was completed on laboratory organisms, which may not react to stimuli the same way as individuals in their natural habitat.
- The study was a computer simulation of a model system, which may have missed a key factor present in the real world.
- The study only considered a small subpopulation of all the individuals possible.

Any of these might mean that the Answers provided in the study might not apply more broadly outside of the specific study system in the paper.

Question 3: Is it clear to you that the experiment designed will collect data that allows the researchers to answer the research question?

Try writing down how each of the variables identified in the research question have been measured/quantified. Are these good proxies for the key parts of the research question? Remember, it's often not possible to directly measure or observe the actual factor or force we are interested in, so the researchers must have chosen something to measure that is possible/easier.

This is part of deciding whether the methods used are appropriate. Particularly by considering whether there are different methods of measuring the key variables or any additional factors that needed to be controlled for. It is possible that the things that have been measured in the paper are not actually good representations of the key factors identified in the research question.

For example, in a hypothetical paper investigating whether natural foraging opportunities improve the stress of captive lemurs, the researchers may have chosen to create a welfare scale based on the number of times lemurs exhibit stereotypical behaviours, which is an industry standard method. However, animal behaviour is an indirect measure of stress. A better study might also measure serum cortisol levels as a more direct measure of stress and check that there is a correlation between the behavioural and physiological measures.

Alternatively, consider a study that is making a claim about the presence of a functional protein, but only measured the mRNA presence. Could they have measured the presence of the protein more directly, or even measured its activity levels?

Question 4: What could the researchers have done to provide a clearer or more broadly applicable answer to the research question?

The authors may have already suggested this in the Discussion, either as a specific paragraph or spread throughout. You can consider whether you agree with their suggestions, or whether you would propose a different study based on your evaluation of the methods and data. A helpful way to think about this is to consider what the study has not measured (e.g. point on protein activity levels above).

This is likely to be a question asked in any tutorials where the paper has been given as reading for a critical discussion. It's also something you should include in the Discussion section in your practical report assignments (Chapter 8, section 8.3). It might be part of the concluding statements in your own coursework where you outline the future of a subject.

Question 5: Are your interpretations of the patterns in the data the same as the study authors?

Most of the Figures and/or Tables presenting data will be in the Results section. Try writing a sentence to explain what each of the Figures and/or Tables actually show and then considering whether these align with the answers to the research question you identified earlier in the process Key Finding of the study that you identified earlier. This is a great way to develop some data literacy, the more you do it the better you will get. Reading graphs is a bit like learning a language and that requires practice.

Every scientific paper is fundamentally just the authors' interpretation of the data. Different individuals may interpret data differently. Have a look at the data and see if you reach the same conclusions as the authors about the relationship between the variables measured. For example, is there a significant trend in the data but the actual effect of the independent variable is very small (effective size). Or is the data very messy with large standard errors, which suggests there are other variables that are important?

[Advanced] Question 6: Have the researchers appropriately analysed their data?

Near the end of the Methods, there will be a statement about what analyses the authors have performed and how they have managed their data. This also usually contains a short explanation of why the authors think this is the most appropriate data analysis method. We have marked this as advanced because the difficulty of the statistics is probably beyond what you have learned on the course (but not necessarily). However, as with the point above, the best way to develop an understanding of statistics is practice.

There is no single correct way to analyse a dataset, and the authors will have made choices based on what knowledge and resources they have access to. Try to consider whether there are other ways to analyse the data that may have been more appropriate.

For example:

- Has the author used parametric statistics on a non-parametric data set?
- Does the study state a sampling effort of 'n' samples, but has in fact collected multiple samples from each individual without correction within the statistics (which is known as pseudoreplication), which would mean the actual study sample is much smaller.

Finally, does the study provide good evidence to support its key findings?

Considering everything you have written for the questions in the rest of this table, do you think the Key Finding being proposed by the authors (which you wrote down earlier) is well supported by evidence within the paper?

This is the statement that you are likely to use in your actual assignment. Remember that we cite scientific papers in order to support specific points in our own work (see Chapter 8, section 8.1.4). You can then follow up the citation with an evaluation of how good that paper is as evidence to support the specific point (as well as evidence of your critical thinking for the person marking your work). For example, picking the lemur study back up, we might say that there is evidence that natural feeding can improve animal welfare, but that the evidence is limited to primate species and not representative of the wider animal kingdom. Or that evidence is overly reliant on behaviour observations and would benefit from more research focusing on physiological measurements of welfare such as well researched stress indicators like serum cortisol.

12.2.5 Critical synthesis of multiple perspectives (does everyone agree?)

A key outcome for many Bioscience degrees is a recognition that scientific knowledge is contested. Science is based on a continuous process of empirical testing of theories and hypotheses that can only be supported and never 'proven'. This often means that as one investigates topics of increasing complexity, there is no single correct answer. As a result of this, for most of your assignments, you will need to consider multiple perspectives on the topic or multiple potential answers to the question. A key purpose of most assignments, by the time you reach your final year of undergraduate study (and certainly the postgraduate level), is to read and evaluate the evidence for different positions on a topic and then synthesise a conclusion based on your independent research of the topic. Below, we outline questions to aid critical synthesis as an extension of the critical thinking process begun in section 12.2.3.

Question 1: What answers to the research questions are the authors expecting they might find?

This is one of the main purposes of the Introduction. Throughout the Introduction the authors will provide an overview of other studies that have investigated the same question and what answers they found.

   Knowing what the different perspectives and theories in the field will really help you with your independent investigation of the subject. A good paper will set up multiple equally plausible predictions and then use the rest of the paper to establish which is in fact correct in the specific context of the experiment. You can use this section as an opportunity to find other relevant papers to read and gain an overview of the history of the subject (which you can use to answer the next sets of questions).

Question 2: Are there other studies that support the Key Findings and/or Answers proposed in this study?

A great place to start would be reading the papers covered in the Introduction section. They may lead to other papers. For the purposes of this bit, just focus on reading the Abstracts of lots of papers and noting down how they support the Key Finding of the original paper. You could also look at the studies that have cited your original paper since it was published.

Question 3: Are there studies that present different, converse Key Findings and/or Answers to those proposed in this study?

A specific internet-based literature search to find contrasting opinions is a great place to start (see Chapter 9, section 9.4.1 on the inquisitive approach to researching the topic). Again, focus on reading the Abstracts of as many papers as you can as you develop a broad understanding of the subject, and keep notes of what they say. A literature review of the topic may well be the best place to find the various contrasting opinions on the subject.

   For each new paper Abstract you read, write a single sentence that outlines how that paper contributes to a greater understanding of the Key Finding (supporting or contradicting) of the original paper considered so far in this process. For example:

- Does the paper find the same answer to the research question in a different study system?

- Does it propose a different answer to the research question?
- Does it suggest how all of the research on the subject might be applied to solve a major problem or improve our understanding of a specific natural phenomena?

Start the process again (from section 12.2.3) with the most relevant paper you have identified in the synthesis process.

Select a paper that provides a particularly important point for your assignment and repeat the process. You could consider what aspects of your assignment you still do not know the answer to and select a paper that provides an answer to that question. This could be an infinite process, so set a limit, the most important thing is to show the marker that you are aware of the key perspectives on the subject and that you have considered evidence that supports each of them.

12.2.6 Evidencing critical thinking in your assessment

For most of your assignments, you will need to show a balance of critical evaluation and critical synthesis. Evaluation shows that you are able to critique specific pieces of information in detail with a depth of understanding of the subject. Synthesis shows that you have read a range of perspectives in your independent investigation and have developed a broad understanding of the subject. Both of these aspects are required to have truly evidenced critical thinking in your work. Try to focus on evaluating key papers that form pillars of your argument.

It is worth noting that you will be unlikely to be expected to scrutinise each paper you read in detail, which would take too long and mean that you probably will not read enough papers to evidence breadth of understanding. The process in this section has been designed to help ensure your assignments have a good balance of both critical evaluation and critical synthesis. You can use this process combined with the assignment planning guidance in Chapter 7 to efficiently create assignments that evidence critical thinking.

Make sure to keep notes of your reading and to reference everything correctly (Chapter 9)!

Try this—Critically evaluate a research paper

Find one of the research papers that you used for a recent assignment and work through the questions outlined in this section to see how you would rate the quality of the content and presentation. You might also like to have a look at some critical reviews of news articles, which pick up on biomedical stories and evaluate their reliability. For example, 'Behind the Headlines' is a resource published by the NHS that comments on the way in which biomedical science is reported in the media.

 Chapter summary

This chapter has focused on two important 21st Century skills that you will need to develop during your degree: collaboration and critical thinking. These are important skills that require knowledge and practice of effective techniques to evidence them effectively. This chapter has been designed to give you some foundational tips, techniques, and processes to help you actively develop, apply, and evidence collaborative and critical thinking skills in your university assignments (and beyond).

Effective collaboration requires the active creation of a professional environment by collectively agreeing ground rules, working respectfully and professionally, agreeing how work will be shared, and how the team will communicate. Make sure that a proper record is kept of what tasks have been agreed, who will do them (work to each person's strengths) and by when; this will help keep track of how the project is developing, and whether any additional actions are required. Most importantly, you must keep meeting regularly and checking progress against your timeline, especially if the project is running over quite a long period.

Critical thinking is the art of evaluating different sources of evidence and weighing them up to draw your own conclusions. This important skill applies equally to reviewing scientific papers as it does to evaluating information in your personal life. This chapter has provided a structured inquisitive approach, with specific questions you can answer that support effective reading of complex scientific literature and thinking critically about what you read. Particularly, we have provided guidance on answering the following questions quickly: is the source appropriate, what are the key points being made, is the source good science [critical evaluation], and does everyone agree with the source [critical synthesis]?

Effectively developing and evidencing both of these skills will be important for your university assessment, especially if you want to get higher grades. We have put them in this chapter separately from the rest of the assessment process as the exact point at which they are needed varies a lot from degree to degree. Remember these skills require practice, so please return to this chapter each time you need to collaborate or think critically and ensure that you build on how you did last time.

 ## References

Thanks to Dr. Katharine Hubbard from the School of Natural Sciences, University of Hull, for their valuable discussion of the critical reading of scientific papers and the Hubbard et al., (2022) study on this subject.

Hubbard, K. E., Dunbar, S. D., Peasland, E. L., Poon, J., and Solly, J. E., 2022. How do readers at different career stages approach reading a scientific research paper? A case study in the biological sciences. *International Journal of Science Education, Part B, 12*(4), pp.328–344.

Stage 3

....

Preparing for your future

....

This final section is comprised of just two chapters:

- Chapter 13, 'Making the most of feedback', and
- Chapter 14, 'Making yourself employable'

Just because these chapters are at the end of the book, it doesn't mean that they're less important, it just makes more sense for the sequencing to put them here. However, both topics can help you throughout your degree and beyond.

Constructive feedback, whilst sometimes difficult to receive, is invaluable to improving performance, but only if you reflect and act on it, and you are the only one who can do this. Similarly, making yourself employable isn't something that is done to you, it needs to be something you actively engage in throughout your degree (indeed, throughout your working life).

We said in the introduction to the book that good study and communication skills aren't only useful for studying for your course; the kinds of skills you need to develop to do well in your chosen field of study have many parallels with the kinds of skills that will be useful throughout your career. We pointed out that being able to research a subject, construct an argument, write a report, present information, manage your time, and plan your own development are all skills that are highly sought after by most, if not all, employers. So, in many ways, you've already started preparing for your future, it's just that you may not have recognised that the skills you were developing to succeed in your studies will also be useful in the world of work.

13 Making the most of feedback

→ Introduction

So, the essay you worked so hard on has been marked and returned to you. You take a quick look at the mark: it's quite good and about what you expected, so you drop the essay into a folder and don't look at it again. You have just done what many students do when they get a mark that was about what they expected or maybe a bit better: you feel quite happy but you have also just denied yourself an important opportunity to improve your future performance. You got quite a good mark but why wasn't it *really* good? What could you have done better? How can you get a *really* good mark next time?

This is where feedback, and how you use it, is a vital component in the foundational skill of developing yourself. This also reflects a common difference between your experience at university and that at school or college. You may have been used to drafting a piece of work, receiving feedback on it from your teacher and then redrafting it before finally submitting it. At university, the common practice for both formative and summative assessments is that you prepare the work and submit it. The feedback will come with the return of the piece of work accompanied, in the case of summative assessment, with the marks awarded.

Therefore, you need to get into the habit of using the feedback from one piece of work to improve on the next one. It is all too easy, however, to make excuses for not doing anything: 'I'm too busy at the moment, but will look at it when I get the chance'; 'I did the piece of work so long ago, there's no point in going back over it now'; 'We've finished that module, so there won't be anything useful anyway'; 'I never understand/can't read what Dr Bloggs writes, so there's no point in looking at it'. No doubt you can come up with quite a few more seemingly plausible reasons for not bothering to revisit the piece of work, maybe not even bothering to collect it in the first place. However, whatever form the feedback takes, even if it is really limited and you only have the bare marks, you should always make the effort to review the feedback and use of it to improve your future work.

You should also be willing to discuss the feedback. For example, if there are aspects that you don't understand, then ask the person who marked the assignment if they can help you. Also, make use of sessions such as personal tutorials to talk through the feedback you are getting. These can be very helpful because they allow you and your tutor to bring together feedback from different assignments and see whether there are any common themes which you could address (see section 13.4 for more details).

13.1 **What is feedback?**

Ideally, feedback should serve three key functions:

1. Provide you with a clear indication of how well you are progressing in your work in relation to the standards established by your university.

2. Give guidance as to how you can improve for when you come to prepare the next piece of work.

3. Flag up things you are doing well and should therefore continue doing.

At its most basic level, then, feedback may take the form of the marks you get for the piece of work: these marks give you an overall indication of how well you are doing. In terms of specific guidance, marks on their own are clearly not very helpful; nonetheless, you can still use them to help you improve (see section 13.3 *How to make use of feedback*).

One level up from this in terms of usefulness is feedback that indicates what you got wrong or did badly. This is probably the most common form of feedback and it may take the form of specific comments such as: 'you did not discuss the role of . . .' or even 'this is wrong'. The marker may also include more generic comments such as 'the essay was poorly structured' or 'not enough detail'. These types of comment are very common, in part because they are usually the easiest type of comment for a marker to make. By pointing out what is wrong, the marker is providing limited guidance as to areas for improvement, but it is often not clear exactly what you should do to improve, particularly in the case of the generic comments that may be quite vague. In this context, therefore, the best feedback is that which identifies the error and then provides guidance as to how to overcome it. This may take the form of full examples, or perhaps pointing you in the direction of specific resources that can help you:

> 'Make sure that you use the appropriate referencing style for your essays—look at the section on referencing in the course study guide.'
> 'You have misunderstood the mechanism of oxygen transport—see pages 32–35 in your physiology course text.'

Less common, but also very useful, is feedback that identifies positive features of your work. Here the marker not only provides encouragement 'this was a good essay' but also identifies specific aspects of the work that were well done:

> 'This was a very good essay and I particularly liked your use of diagrams to support the arguments you were making.'

Again, the value of the feedback is greatly increased where examples are given so that you are clear about what things you should carry on doing. Inevitably, the quality and level of detail of the feedback you receive will vary. Indeed, recent National Student Surveys (NSS) have consistently identified feedback as one of the aspects of higher education that students find least satisfactory. However, it must be remembered that feedback is a two-way process and you will only benefit from it if you engage with the process. In this way, even if the feedback you receive is not always very detailed or is late in coming, you can still be able to identify ways of improving what you do.

13.2 When do you get feedback?

Many students only think of feedback in terms of written comments on coursework assignments. Indeed, this is the most obvious and one of the most common forms of feedback. However, feedback comes in a variety of forms and, unless you are ready to recognise when you are receiving feedback, you will be missing out on valuable learning opportunities. Let's now consider the range of different feedback types that may be available to you.

13.2.1 Written comments on written assignments

As stated earlier, written comments on written assignments constitute the most obvious form of feedback and are probably the most common form of formal feedback. These types of comments usually take two forms:

1. Generic comments, usually written at the end of the piece of work, or on a separate cover sheet. These comments tend to relate to the overall style and quality of the assignment.

2. Specific comments which usually take the form of annotations made on the script itself. These may relate to stylistic issues, such as paragraph structure or referencing, or they may be subject specific, such as the identification of a factual error or the omission of a point in the argument.

13.2.2 Comments on oral or poster presentations

These feedback comments may be given in the form of written comments on a mark sheet, but they are often given just as oral feedback following the presentation or discussion of the poster. Such feedback can be very useful because it is the most immediate form of feedback, being given within a few minutes of the delivery of the piece of work. However, oral feedback is also transient: if you don't write the comments down you are likely to forget them very quickly and then you won't derive any long-term benefit from them.

13.2.3 Comments during academic tutorials

Tutorials are often a source of very useful discussion about specific topics in the curriculum (see Chapter 3, section 3.2 *Tutorials*). However, as with the verbal feedback on oral presentations, the points made by a tutor will be transient and quickly forgotten unless you note them down. Tutorials provide valuable opportunities for discussion and for you to raise questions if there are points that you don't understand. If you have prepared well for the tutorial you are also in the best position to benefit from the feedback from your tutor.

As well as noting down the comments made about your own contributions, you should also be prepared to note down comments made about the contributions of your fellow students as these are also likely to be of relevance. Tutors probably consider that they are providing you with valuable feedback during tutorials, but students often do not recognise it as such and miss out on the opportunity to learn.

13.2.4 Comments during practical classes

Practical classes offer another very useful way in which students can engage in discussion with tutors or demonstrators. As a consequence, they can be a rich source of feedback, for example, in relation to experimental technique, analysis, and presentation of results, or in understanding of the topic itself. Again, however, because it is spoken, much of this feedback is transient and it is very important that you note the comments before they are forgotten. Also, remember to take notes during any briefing session before or after the practical class.

13.3 How to make use of feedback

If you are going to benefit from feedback that is provided, then you must take steps to engage with it as an active process. This doesn't just mean looking briefly at the comments. Engagement requires a detailed reflection of how the comments relate to the piece of work you have submitted and consideration of how you should change your approach to future pieces of work.

There is often a perception that if feedback is not provided very quickly, then it is of no value. The usual reasons for taking this view are:

- that it is several weeks since you did a piece of work and you can no longer remember much about it;
- that you have already had to submit another assignment before getting the feedback from the previous one;
- that the module for which you did the piece of work has now finished and therefore the feedback no longer appears relevant.

These reasons, while initially understandable, are misguided even though you may feel frustrated by the delays. It is worth noting that most universities publish the expected turn-round times for marking and if there are unexplained delays, you have the right to query them directly or through your course representative.

Even if the feedback is slow in coming, there are always useful pointers that you can take from it, though it may be that sometimes it requires more work on your part to extract the maximum value.

For the most part, in this section, we will focus on making use of written feedback since this is the most common form of formal feedback on coursework; however, much of the guidance also holds true for oral feedback. The difference between the two is that in the latter case, it is up to you to make sure that you make a record of the feedback, for example, by making notes of discussions in tutorials or during practical classes.

13.3.1 Engaging with written feedback

If you have a cover sheet, as is used by many universities, or if you have been given summary written comments at the end of the assignment, then read these carefully before doing anything else. Divide the comments into two groups.

1. Generic comments: ideally these take the form of 'feed forward' comments such as identification of broad areas for improvement. This could include how to improve your overall writing style, to reference better, or to improve the way you draw graphs and present data. There may also be positive comments, identifying things that you have done particularly well and should carry on doing in future work.
2. Subject-specific comments: these will normally be related to factual information such as the identification of factual errors or omissions. Again, markers should also flag up good points, so you get a feel for what you are doing well and for what needs improving.

Having categorised the comments, read through the piece of work carefully; after all it may be some time since you wrote it. As you read through, bear in mind the comments that were made and try to be objective in your evaluation. Also, make sure that you can identify what aspects of the essay or report the comments relate to. You should also note any annotations on the script. With the generic comments, make sure you understand how the comments apply to your work and how you can improve what you have done. Make a note for future reference so that you can carry them forward to your next assignment: although the next assignment may well be for a different module, generic issues such as referencing style or essay structure will still be relevant. Likewise, make a note of aspects of your work that were highlighted as being particularly good. With subject-specific comments, check that you understand any factual errors or omissions. Make a note to go alongside your lecture notes to make sure that the point is clear for revision purposes.

Look at your marks and check them against any marking criteria that you might have been given for the assignment or against any generic criteria that your department might have published. Make sure that you appreciate how your work fits against the criteria and why it falls into the specific mark band that was allocated. Again, if you don't understand, then ask.

For example, if you obtained a mark of 55%, the marking scheme might state that the work has the following characteristics:

> The essay is well organised, displaying understanding of the main issues. There may be a few, minor errors or poorly expressed ideas. There is a significant dependence on lecture notes and/or textbook material.

As you read through your work, make sure you can identify why the marker placed it in this category: were there some minor errors? Did you mainly use lecture notes or textbooks rather than research literature (see Chapter 9 *Researching your coursework topic*)? Think about how you could do it better next time.

Feedback may not always be clear! For example, if you don't understand how to improve even though the marker has highlighted a weakness. You might not understand what the marker means by some comments. If you are not sure what to do, then do ask your tutor to explain in more detail how you can improve: arrange a meeting with the tutor and take the work along with you so you can discuss the strengths and weaknesses in detail.

Make sure that you take the feedback forward and use it to improve your next piece of work and your understanding of the subject: when you are preparing for your next assignment, remember to read back through the notes you have made on previous pieces of feedback so they are fresh in your mind. When you have taken the points forward, also check the feedback on the next piece of work to ensure that you have indeed improved.

Try this—Swap your feedback comments

Try this feedback activity with a friend. Take the most recent piece of work you have had returned and swap the feedback that you received. Identify the skills that your friend needs to improve on and encourage them to note down how they will improve them. Ask your friend to identify the skills that you need to develop and make a note of how you will improve them. Then, make sure that you refer back to these notes for the next assignment.

13.3.2 Engaging with oral feedback

As indicated earlier, this may take various forms, for example in the laboratory, a tutorial, or as feedback on a presentation. The danger with any of these forms of feedback is that they are very transient and easily forgotten, therefore you must be sure to take notes so that you can remember the specific points. The principles, though, are the same as for written feedback: divide the feedback up into generic and specific aspects, and make sure that you understand what the issues are and how to take them forward.

Feedback, then, is one of the keystones of good learning, but you can't treat it as a passive process, otherwise you will not progress and you will make the same mistakes again and again. Engagement is all-important.

13.4 Personal tutorials

Personal tutorials may cover a range of topics. In relation to the themes of this book, the most important aspects are likely to be topics such as your academic progress, module choices, personal development planning, and thoughts on careers. Again, if you are going to benefit from the opportunity then you must prepare in advance and be ready to discuss these issues openly.

Personal tutorials are a very useful opportunity to talk about the feedback you have received on different pieces of work.

Why is this important? Firstly, it provides a good opportunity for discussing your progress and the feedback you have received across several pieces of work: for example, are your marks improving? Secondly, it is important to remember that, at university, it is common practice for assignments to be marked anonymously and many markers may only mark one piece of work you have submitted. If the marker makes a comment on your work, for example, about paragraph structure or use of references, they won't know whether they are the first person to make that comment or if several other markers have also said the same thing regarding other pieces of work. Only you will be in a position to bring those different pieces of feedback together, and so it is useful to be able to discuss this with a personal tutor or adviser. Again, if there are aspects that you don't understand or feel you need help with, do discuss them with your tutor so they can advise you, or refer you to one of the sources of support offered by the university.

> ### Try this—Common feedback comments
>
> Collect together the last three or four pieces of work which you have had marked and read through the feedback. Make a note of any comments that are made more than once. These are clearly areas for you to focus on in future and possibly discuss with your tutor.

To discuss your proposed module choices for a forthcoming semester or academic year, you need to have read about the range of choices that are open to you. This will include checking whether taking the modules requires you to have taken specific pre-requisite modules in the current year of study. You also need to have thought carefully about where your strengths and interests lie, and indeed, if you have any particularly weak areas of the subject that are best avoided. In the context of your development, you should also be thinking about whether you would benefit from further training to develop your skills. For example, if the modules are assessed by poster presentations and you are not sure how to go about preparing posters, then you should read Chapter 10, section 10.2 *Creating assignments focused on visual communication*, and also consider seeking guidance from your tutors or perhaps from the university's learning support centre, or equivalent. Finally, you should also be thinking carefully about where you think studying those modules might take you in career terms, with regards not only to the subject content, but also the specific skills that you will be developing (see Chapter 14 *Making yourself employable*).

Try this—Preparing for a personal tutorial

This is a useful exercise, even if you are not scheduled to be meeting with your tutor in the near future, as it gives you the opportunity to take a step back and reflect on where you are in your programme. Try writing a sentence or two in answer to each question:

- Am I enjoying the course or are there some aspects that I enjoy more than others?
- How am I coping with the different modules and how am I dividing my time between them?
- Are my marks as good as I expect/would like?
- Have I read through the feedback I have received on my work and have I acted on it?
- Are there any specific aspects that I am struggling with—this could be specific topics or activities such as essay writing? If so, do I know where to get help?
- What are my plans for the next set of modules or for when I graduate? Am I taking the right modules at the moment to help me transfer to the next stage?
- Are there any personal issues I want to discuss with my tutor or with other university support services?

Do try this just before a tutorial, but also around the mid-point of the term; it is good practice to reflect on how things are going.

13.4.1 Contributing to personal tutorials

With personal tutorials, you can only expect to benefit if you engage actively in the process and, in these cases, are prepared to be open in terms of your views and aspirations, as well as being realistic in terms of your abilities. This means that you may need to open up discussion topics and explain your background to your tutor, rather than waiting for the topics to be raised. Despite the best of intentions, your tutor may not be able to guess what is on your mind or worrying you at the moment. You should also recognise that your tutor may not have all the answers, although they should be in a position to be able to advise you on where to seek further guidance.

It is important to remember that your personal tutor is there to try to help you. Often students are anxious that if they raise issues of concern with their personal tutor, it will affect how the department views them. This should not be the case under any circumstance. Although the personal tutor is not bound by a confidentiality clause as are, for example, GPs or counsellors, they have a duty to be discreet about things they are told and not to be judgemental.

If you have issues that concern you or you want to seek advice, don't be afraid to make an appointment to see your tutor. Some departments operate an 'office hours' policy, whereby tutors will always be available at specific times during the week. Other departments may have an 'open door policy' where you can go to see a member of staff at any time. Either way, it is often a good idea to contact your tutor first to make sure you can see them at a specific time that fits in with your timetable. Scheduling a meeting also means that your tutor should have time for a proper conversation.

Chapter summary

Figure 13.1 shows a flow chart of the processes of using feedback. The most important aspect of this is that you need to think about the feedback, see how it relates to your work, and then to see how you can use it in future to improve your performance.

Figure 13.1 Engaging with feedback

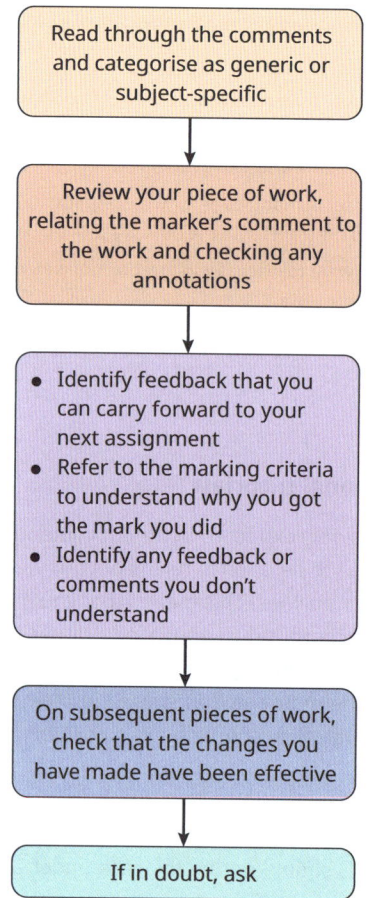

14 Making yourself employable

→ Introduction

We considered putting this chapter at the beginning because thinking about employability at the end of your degree is far too late. However, we identified making yourself employable in the opening chapter as a set of skills that would be addressed throughout the book and have highlighted the employability applications of study and communication skills at relevant points, so hopefully you have got the message that making yourself employable is critical. Having said that, if you are reading this chapter in your final year do not despair! This chapter can still help you to make the most of the time you have remaining at university to make yourself employable.

The wording of the title is important too—I'm afraid we can't guarantee that following the advice in this book will get you into employment (though we certainly hope it will), but we are confident that if you follow the advice, it will make you employable. And when you think about it, being employable is, in many ways, more important than being employed: jobs no longer last a lifetime, and when you find yourself looking for work, it's employability that counts. Peter Hawkins, in his book *The Art of Building Windmills* (1999), put it like this: 'To be employed is to be at risk, to be employable is to be secure'. So regardless of what the current job market happens to be doing, or which jobs artificial intelligence might be coming for next, the important thing is to make yourself employable so you can increase your chances of being employed whatever the circumstances. This chapter will highlight the importance of getting a good degree, engaging early with employability, gaining experience, making contacts, and selling yourself in the context of the selection process.

14.1 Get a good degree

This book is all about helping you to get a good degree, and if you've invested a lot of time, effort, and money on your studies you'll want to get a good result. The classification of your degree is one of the things many employers still look at in their selection processes. The larger graduate employers may receive hundreds of applications for each position so they will want to filter out applicants based on easily quantifiable criteria, such as degree classification and even UCAS tariff points. This means that for some graduate recruiters (but importantly not all) if you don't get a 2:1 degree you'll not get past the application form stage. However, this trend is declining. The Institute of Student Employers track academic requirements stipulated by employers in their annual Student Recruitment Survey. Between 2013 and 2023 the number of employers:

- requiring a 2:1 or above reduced from 76% to 48%;
- requiring minimum A-level grades dropped from 40% to 9%;
- specifying no minimum academic requirements increased from 6% to 22%.

If you get a 2:2 (or think you are heading for one), you will need to research which employers are happy to accept a 2:2 or don't use degree classification in their screening process (your university careers service will be able to help you with this) and then focus on those. Bear in mind, though, that some employers who have a 2:1 as a minimum criterion may choose to ring-fence a handful of places for graduates who have a 2:2 but can demonstrate outstanding strengths in other areas (for example skills, experience, or attributes).

14.2 Engage early

As we said in the introduction: trying to make yourself employable at the end of your degree is leaving things too late: you have to engage early on. Admittedly, thinking about your CV is probably not at the forefront of many first years' minds during welcome week, but once you've had a few weeks to get settled, it's a good idea to look a bit further ahead. This doesn't necessarily mean looking for internships to apply for after Christmas, but it should include doing some of the following:

- getting involved in clubs and societies;
- finding part-time work;
- looking out for opportunities to get involved in student representation (e.g. becoming a course rep);
- starting to think what areas of employment interest you.

None of these things are *directly* about employability, and you might do them for reasons other than employability (trying new things, meeting people, earning money, etc.), but they all have employability benefit. By doing these things, you'll be gaining useful skills and experience that will make both thinking about, and gaining, graduate employment at a later stage considerably easier. Of course, if you're coming into your second/penultimate or final year, there will be less time left for experimenting and you'll need to be more focused on the specific skills and experience related to the areas of employment that interest you.

The reason you need to start early is because a good degree is not the only thing that employers look for. In fact, in many ways, employers take the degree as a 'given' and it's everything else that really counts (especially as more graduate employers drop their academic requirements). This is less the case if you are using your degree to get into an industry that requires very specific academic qualifications, or want to pursue a higher degree (Masters or PhD), but, regardless of which industry you are targeting, it's always the additional things that make you stand out. Employers will be looking for a good degree plus:

- transferable skills (team-working, communication, project planning, problem solving etc.);
- relevant experience (dealing with customers, positions of responsibility, etc.);
- personal attributes (motivated, down-to-earth, get on with others easily, etc.).

and all of these will need to be effectively communicated through the appropriate part of the selection process, whether that be by CV and covering letter, application form, assessment centre, or interview.

A useful maxim is this: figure out where you want to be and then work backwards; what is it you want to do and what do you need to get there? For example, if you want to be a teacher you will need experience of working with children and have had several placement days in a school. Couple this with the fact that applications for teacher training need to be submitted 6–12 months in advance (depending on which route you choose) and you can see why you need to start early. For teaching, ideally you need to begin working on your relevant experience early in your penultimate year if you want to get on a teacher training course when you graduate. Likewise, if you want to use your degree to work in the pharmaceutical industry, then you would be advised both to aim for a laboratory-based project and to obtain some additional laboratory experience during your undergraduate studies. This experience could take the form of a placement in a university or other research lab during the vacations, or as a placement year in industry. So, if you want to follow this route, you need to be planning early: probably at the start of your second year. You also need to choose your final year project carefully, aiming for one that will give you more relevant laboratory experience.

If 'figure out where you want to be and then work backwards' seems a bit overwhelming, it's helpful to remember that this isn't a decision about the career direction for the rest of your life; just think of it as your next step. However, many (perhaps most) students genuinely don't know what they want to do, if this is you, you need to start a bit further back.

14.2.1 Know yourself

There are many good books and websites available on career planning, so we are not trying to cover everything here—just the basics. First, you need to understand a little bit about yourself. It's surprising how many people miss this first stage out, and then wonder why they can't decide what to do or don't feel suited to what they end up doing. But it's really not that complicated; you just need to reflect on a number of questions, such as:

- What is most important to me?
- What are my core values?
- What do I want and need from a job?
- What motivates and interests me?
- What strengths, skills, and experience do I have?

It is answers to these questions, and questions like them, that will help you to know yourself. And knowing yourself will help you to decide the kind of thing you want to do. Do you want to be a research scientist in a pharmaceutical company? An academic? A consultant in a professional services firm? A journalist? A business developer in a start-up organisation? Your own boss? The possibilities are endless! It does take a bit of time and effort (which is why it's so often missed out) but it is a vital starting point. Try talking the questions over with people who know you well, such as your friends or family—not that you have to do what they think you should do, but it will provide a useful perspective.

Employers want you to know yourself too; they are trying to attract not just bright, competent graduates but bright, competent graduates who have an interest in, and passion for, their business or profession. Employers spend significant amounts of time and money looking for, and then training, their graduates; they don't want to recruit someone who, three months into the job, decides they don't want to be an accountant after all and wants to do a PhD instead. Therefore, their selection processes will include elements that try to identify your motivation for wanting to join them (and them in particular, not their sector or profession in general): if you can't convince them that you are motivated, they will filter you out of the process, however bright and competent you may be.

14.2.2 Explore your options

Once you understand a bit about yourself, and in particular, what motivates and inter- ests you, you are in a much better position to explore your options. Otherwise, it's a bit like walking into a huge library thinking: 'I'd like a book'! You need to have narrowed down your preferences for your search not to be futile or random (and probably quite demotivating).

A useful starting point is to think about whether you want to use your Biosciences degree in a specific or a generic sense. Do you want to become, for example, a researcher, embry- ologist, botanist, geneticist, or science teacher? Or are you more interested in a graduate job that requires a good degree generally rather than a Biosciences degree in particular? It comes back to thinking about where you want to be and working backwards.

Often, if you want to use your degree in a specific sense, you will need to undertake further study, but you need to be sure of this before you embark upon a higher degree or other form of additional training. You will need to do your research to find out exactly what kind of further study the employers or professions you are interested in require. Do they require a Masters or a PhD or some other postgraduate qualification? Are there other ways in aside from the postgraduate qualification route? Is the institution that you are thinking of studying at well regarded by employers? What track record does that institution have of getting its postgraduates into graduate jobs? Postgraduate study is a big investment of time and money and it's a bad idea to use it as a gap-filler simply because you don't know what you want to do at the end of your undergraduate degree. Otherwise, you might find yourself with more letters after your name, a bigger debt, and still no clearer idea as to what you want to do.

BOX 14.1 Postgraduate study: is it for you?

As we have already indicated, you will already have invested a significant amount of time and money in your current degree programme and postgraduate training represents further investment. It is important, therefore, that you are very clear in your own mind as to why you are embarking on another qualification. So, the first step is to explore exactly what the employers are looking for.

Postgraduate study can take a number of forms, the most common of which are:

- vocational conversion courses;
- taught Masters programmes;
- postgraduate research degrees.

Vocational conversion courses, as their names suggest, allow you to change direction from your Biosciences degree to another professional route. Among the most common such programmes are conversion courses that are linked to specific professions, for example, in law, finance, and areas such as psychology or social work. Some of these programmes may carry academic awards such as a postgraduate certificate or diploma or a Masters degree. The most important thing, though, if you are going down this route, is to make sure that the course is recognised by the relevant professional body and so allows you an entry into that profession. Your careers service will be able to advise you regarding the range of courses and the different forms of recognition.

Taught Masters programmes are usually one-year courses (if taken full-time) which will allow you to gain a higher academic award than your first degree. There is a huge variety of such awards available in a wide range of specialisms. Some may carry bursaries, but these are relatively scarce. The likelihood, therefore, is that you will have to find the money for the course and your living expenses, or (more likely) apply for a postgraduate loan (www.gov.uk/masters-loan/apply). You need to do your research very carefully and ask yourself why you are thinking of doing this course and where it could take you: for example, look at the employment profiles of the graduates from the programmes you are considering and, again, talk to your careers service. It's a good idea to try to gain some experience before you apply, to test out your ideas. For example, if you are interested in an MSc in Ecology, you should look for paid or voluntary opportunities in an ecology-related role; this would both test out your ideas and make you more eligible for the course. Just gaining another qualification, for its own sake, is probably not a sensible strategy.

Research degrees also take different forms: at the highest level is the doctorate (PhD or DPhil) which typically lasts 3–4 years and is predominantly based around a specific research project, though increasingly they also involve training in a range of research and generic skills. There are also MRes and MPhil opportunities that offer a shorter period of research. The MRes usually allows you to undertake more than one short research project, so you can explore which area you might want to specialise in and, indeed, whether research is for you. There are, quite often, various awards available to support research students undertaking doctorates, for example, studentships funded by the research councils. Again, though, you need to consider very carefully why you want to follow this route and you would be strongly advised to gain some practical laboratory experience first, both to strengthen your CV and to help confirm in your mind that this is what you really want to do.

If you want to use your degree in a generic sense, this too is going to require some research. Many, if not most, graduate employers accept students from all disciplines. They are interested in good qualifications, relevant work experience, and transferable skills, but exactly what subject the qualifications are in is usually secondary to the work experience and skills. In addition to thinking about what kind of role you are best suited to (knowing yourself), it is also sensible to think about the current state of the industry or sector you are interested in. For instance, are the police currently recruiting? Is the pharmaceutical industry in the UK growing or shrinking? Are investment banks currently taking on graduates? Are there local companies that are expanding? Will this job exist in five years' time (see Box 14.2: How will technology change the job market?)? Knowing the answers to these kinds of questions will give you an indication of how competitive your quest for a graduate job will be. That's not to say you shouldn't aim for the most competitive roles (it's about making yourself as eligible as possible, not leaving it to chance) but it is a good idea to have a back-up plan should your first choice role not come to fruition.

Your university careers service will be able to help you explore your options. If you've never been to see anyone in a careers service before, don't let that put you off. They are there to help you to think through exactly these kinds of issues. If you do some preparation beforehand, reflecting on specific questions you want answers to or specific things you want help with, it will make the time you spend with an adviser all the more useful.

BOX 14.2 How will technology change the job market?

This is a huge topic and one we can't possibly do justice here, but it is important to raise nonetheless. Technological advancements have always changed society, and that includes changing work. Periodically, however, changes come along that revolutionise work: the printing press; the steam engine; the mechanisation of agriculture; the desktop computer; increasingly sophisticated automation. Just think about how you use the checkout in a supermarket now compared to five years ago. Or even more recently big data, artificial intelligence, and machine learning, all of which are having an increasing impact on society and the world of work.

Whilst some might think that these advances are bad news for jobs, the World Economic Forum's *The Future of Jobs Report 2023* predicts the 'impact of most technologies on jobs is expected to be a net positive over the next five years.' The report identifies 'big data analytics, climate change and environmental management technologies, and encryption and cybersecurity' as the 'biggest drivers of job growth'. Technology will also result in some jobs declining, principally clerical or secretarial roles, but they predict that job growth caused by technology will outstrip job losses, hence predicting it will be a net positive effect.

It is also worth highlighting what skills employers are likely to be looking for. The report says that:

'Cognitive skills [the functions your brain uses to think, pay attention, process information, and remember things] are reported to be growing in importance most quickly, reflecting the increasing importance of complex problem-solving in the workplace. Surveyed businesses report creative thinking to be growing in importance slightly more rapidly than analytical thinking. Technology literacy is the third-fastest growing core skill.'

The jobs that will be redefined won't just be low-skilled jobs. In their book, *The future of the professions* (OUP, 2015), Richard Susskind and Daniel Susskind argue that 'increasingly capable systems'—will bring fundamental change in the way that the 'practical expertise' of specialists is made available in society, including doctors, teachers, accountants, architects, consultants, and lawyers.

So what does all this mean for you in terms of making yourself employable?:

1. It's helpful to *understand which jobs are most likely to become automated*; these are the jobs that are the most routine and repetitive (hence the supermarket checkout).

2. Think about *what aspects of employment are likely to be least susceptible to automation*; these are roles that involve genuine creativity, involve building complex relationships with people, or roles that are highly unpredictable.

3. Ensure you remain employable by *cultivating those skills that are likely to be most in demand* (creativity, building complex relationships with people etc.) so that, as the labour market changes, your skills and experience remain well sought after.

4. Finally, improve your technological skills. As artificial intelligence becomes more accessible and sophisticated (ChatGTP, Bard, etc.) and more integrated (Microsoft Copilot, Google Duet), your ability to interact with, and make use of, the technology appropriately and intelligently will become more important.

Whilst this might sound a little daunting, note that it's always been the case that those who keep reflecting on and developing their skills are always going to be more competitive in the job market.

14.3 **Gain experience**

At some stage of a selection process for a graduate job (almost any job) you will be asked to respond to statements or questions like:

- Give an example of when you had to manage a challenging situation.
- Describe a situation in which you influenced or motivated people.
- What is your proudest achievement?
- What motivates you?

It's very difficult to provide convincing responses based on academic experience alone. What recruiters will be looking for is a breadth of experience that draws on both curricular and extracurricular activity, not just academic qualifications. Your academic qualifications will only take you so far, and for many roles this is only as far as the first sift of the selection process. The things that will make you stand out are the things that you have in addition to your academic qualifications; your experience.

It's not unusual to hear the following frustration expressed during the search for a job; 'They're asking for work experience—but I can't get work experience without a job, and I can't get a job without work experience!' This is understandable, but wrong. There are lots of ways to gain experience apart from employment, particularly in a university context. In fact, there will be far more opportunities available to you than there is time for you to do them; think of welcome week and all the clubs and societies vying for your attention. There are lots of factors you might consider when deciding what extracurricular activity to get involved in: it might be something that you've never done before that you want to try out; it might be something that you already do that you want to get better at; it might be about meeting new people and making friends. It might be all or none of these reasons, but you should also bear in mind the experience you will gain and the skills you could develop. It doesn't mean you have to be the president of the society or on the committee, but it does mean that you need to be an active member rather than a passive recipient; someone who gets involved rather than just turns up.

Active participation in student clubs and societies isn't the only way to gain useful experience whilst at university: undertaking part-time work, doing an internship, becoming a course rep, learning a language, creating online content, coaching sports, getting involved in volunteering, a year out in industry, a year abroad, starting your own business, or gaining a skills award are all things that will undoubtedly provide you with useful experience. The trick is picking something that you are motivated to do and could be useful in terms of where you want to get to. The earlier you get involved in these kinds of things as a student the better; it will give you a chance to try things out and reflect on what you are, and aren't, suited to. As your ideas of what you want to do become clearer and you narrow down your options, you can begin to get involved in things that are more and more relevant to what you want to do. For example, if you want to become a management consultant, some commercial experience will be helpful; if you want to become a researcher, you will need experience in a lab (see Box 14.1); if you want to get into the media you can join a student media society.

14.3.1 Reflect on your skills

'Experience is not what happens to you; it's what you do with what happens to you' is a quote attributed to Aldous Huxley[1]. It makes an important point because it's not just about having experiences that counts, it's learning and developing through those experiences that's important. Again, as with knowing yourself and exploring your options, this will take a bit of time, but it will be time well spent. You need to think about what you are good at and what you're not so good at. You need to think about how to play to your strengths (the things you're good at) and whether the things that you're not so good at matter or not in terms of the kind of role that you are looking for—and if they do matter then you need to think about how you will develop in these areas.

Again, your university careers service should be able to help with this: they can provide you with diagnostic tests to assess your strengths and weaknesses or skills awards programmes that provide a formal structure to help you reflect on, and develop, your skills.

Try this—Identifying skills developed through problem solving

Ask each one of a group of friends to identify a problem they were faced with and managed to resolve. The problem doesn't just have to be connected to your course; it can be from any area of life. For example, it could be an issue with housemates, something connected to work experience, or any club activities. It may be a problem that affected only you or affected others as well.

As each member of your group describes their problem, make a list of the skills that you think they demonstrated.

14.3.2 Develop your skills

We've been referring to 'skills' throughout this chapter but have yet to say what these skills actually are. Well, this whole book is about study and communication skills and certainly those are included, but graduate employers look for more than just these. They do, however, look for a broadly similar set of skills; these may be referred to by different terms (transferable skills, core competencies, skill sets) but they all mean similar things and they all contain similar elements. Typically, graduate employers will generally look for the following skills:

- communication;
- teamwork and leadership;
- researching and analysing;
- problem solving and decision making;
- initiative and creativity;
- planning and organising;

---

[1] He actually said: 'Experience is not what happens to a man; it is what a man does with what happens to him' (*Texts and Pretexts*, 1932, Chatto & Windus), but this is the given version that has become commonly used.

- learning and development;

- self-management;

- using technology.

It can seem quite daunting to be faced with such a broad range of skills, but this is where you need to think about what you are already good at and what you are less good at and then begin to fill in the gaps step by step. Skills can't be developed in the abstract or in theory; they need to be developed through concrete experience. As we've already identified, opportunities to develop skills through experience abound at university: it's just a case of picking something that motivates and interests you. If it's communications skills that you need to develop, then you need to look for opportunities to present or write; if it's teamwork you need to work on, then put yourself in situations where you have to work in a team to achieve a shared objective; or if it's leadership that you need to evidence, then find opportunities to lead. All of this is obvious, but putting yourself in situations where you are stretched and developed is not necessarily an easy thing to do. But think where it might get you . . .

Try this—Auditing skills levels

For the following skills associated with group work, give yourself a rating and identify an opportunity for further development.

	Novice				Expert
Do you contribute to tutorials, seminars, workshops?	1	2	3	4	5
Are you good at listening as well as contributing?	1	2	3	4	5
Can you assume a variety of roles, e.g. team leader, team player?	1	2	3	4	5
Do you relate well to a diverse range of people?	1	2	3	4	5
Can you give and accept constructive criticism?	1	2	3	4	5

What opportunities might you have in the next few months to practise and develop these skills?

14.3.3 Articulate your skills

Reflecting on, and developing, your skills is important, but if you can't articulate your skills in a way that employers can relate to and understand, the effort may well be wasted. For example, one applicant might describe their part-time bar job on their application form like this:

> 'Worked part-time in Students' Union bar for 12 months.'

Another might describe exactly the same experience like this:

> 'Committed and reliable member of Students' Union bar staff team: worked as part of a team, providing excellent customer service, often under pressure.'

It's not just that one description is shorter and one is longer, the second example actually describes the responsibilities of the role and uses active language to communicate the skills and qualities involved. The candidate in the first description perhaps hasn't reflected on the skills that they've been developing and certainly hasn't articulated them; consequently, they are underselling themself. Try to get into the habit of regularly reflecting on the skills you are using or developing in these kinds of contexts. If your university encourages personal development planning, then there will be resources to help you do this. Your university careers service will also be able to help you to reflect on the skills you are developing and think about how you can articulate them in the kind of language that employers are looking for.

14.4 Make contact

There are far more new graduates each year than there are graduate jobs, but it's important to realise that the graduate programmes of the big employers represent only a fraction of the graduate jobs available. Small or medium-sized enterprises account for the larger proportion of the opportunities available but are often overlooked because they don't generate as much publicity as the graduate programmes of the big brands. This is important to remember when reading statistics on the average numbers of applicants per graduate job, which vary in any given year but are usually quite daunting on the face of it. There are three things to bear in mind when reading such figures:

1. the figures relate mainly to the larger employers and competition amongst smaller employers is much less;

2. the volume of applications is different to the quality of applications: just because there are lots of applications (and so the average number of applicants per graduate job is high) it doesn't mean the applications are any good (you need to make good applications not lots of applications);

3. even in times of very stiff competition, some graduate recruiters will not fill all their places—this is because employers will always prefer to carry vacancies rather than recruit the wrong people.

In a competitive market, it's essential that you give yourself as much of an advantage as possible; and that's not just about the skills and experience you have and how you communicate them but also about the contacts that you make.

14.4.1 Meet employers

One important way that you can stand out to employers is to make them aware of you before you have applied. This might seem unrealistic for the larger employers but it's not necessarily the case; there are many opportunities to meet employers before you even begin the application process, you just need to target them appropriately. Not only is this important to help you stand out, it can also, if approached in the right way, provide you with greater insight into their recruitment and selection process and give you a better idea of what they are looking for in their recruits. This can give you a distinct advantage over other applicants who haven't taken the time to do this.

Your university careers service will be able to tell you about opportunities to meet employers on campus or elsewhere. Some universities are better at attracting employers than others, but all will at least have some employer activity happening. The most common ones are careers fairs and employer presentations, but look out for other opportunities such as employer visits and special events. Opportunities like these aren't just about graduate roles but are often the way to find out more about internships and placement years too.

Before you meet an employer, make sure you do some basic research: look at their website and find out what you can about their graduate schemes and their recruitment and selection process; make the most of face-to-face contact by planning questions that you can't find answers to easily online. This will not only give you information that other applicants may not have, it will also show your knowledge of, and interest in, their organisation.

14.4.2 Use your network

In addition to opportunities to meet employers that are put on for you by your university careers service or by individual employers, you should also think about what other contacts you have that you could make use of. Is there anyone you know who already does what you want to do or is in the industry or profession you are interested in? This could be a direct contact such as a family member or friend or an indirect contact such as a friend-of-a-friend, a member of university staff, or a recent graduate from your course. It might seem tenuous or perhaps a little cheeky, but approaching these kinds of contacts can sometimes make all the difference in your search for information, advice, experience, or even a job.

If you are interested in getting a role in human resources, why not contact your university's Human Resources (HR) department and ask to meet with someone to chat about what they do? Or perhaps your friend's brother works for the big retailer you have an interest in working for: why not drop them an email with a few questions? Or maybe someone from your department has just finished their PhD and gone on to work in the biotechnology sector: why not contact them via one of your lecturers who knows them? Even busy people will often be happy to respond to these kinds of requests, provided they are presented politely and don't look like they are going to take a large amount of time. It's all about being proactive and doggedly pursuing help in a friendly and respectful fashion. You have nothing to lose and potentially a lot to gain.

Think too about how you can create a useful online network. LinkedIn is the most obvious example for professional roles, but other platforms may be appropriate for more niche industries; just look for whatever platforms the companies you are interested in have a presence on. Be careful though, if you choose to connect with a company on a platform that you use for social rather than professional purposes—remember they can see you as well as you can see them. If using LinkedIn, think about connecting not just with the official company page but also with people who work in that company. These people could give you potentially useful insights and help you build a professional network. Cold contacts are less likely to be successful, so search for alumni from your university who work for the company and personalise the invitation to say you would like to connect. To build a network, though, you will first need to fill out your profile information, see section 14.5.1 *CVs*, for help with this (as you can use the information from your CV to populate your profile).

14.4.3 Where to look for jobs

When you are ready to look for vacancies, how do you know where to look (if you are looking for postgraduate courses, see Box 14.3)? It is important to understand that there are advertised vacancies and there are unadvertised vacancies. Estimates vary, but most experts agree that there are more vacancies that are unadvertised than are advertised. Precisely where you look will depend on what it is you are looking for and what type of role: voluntary

---

BOX 14.3 Postgraduate study: where to look

Before embarking on further study you should have a very clear idea of what it is you want to do and why. There are various forms of further study including conversion courses, taught postgraduate programmes (Masters programmes, etc.), and research degrees (see Box 14.1).

If you are looking for a taught programme such as a conversion course or an MSc, then you can search for these online using specific keywords (e.g. 'Law conversion course', or the name of the specific area of Bioscience study, such as 'molecular genetics'). Having drawn up an initial list of potential courses, then you should investigate them in the same way you might have done when choosing your undergraduate programme by looking at:

- the reputation of the institution offering the course;
- the reputation of the course itself and the employment record of the graduates;
- the detailed content of the course;
- the cost of the course and whether there are any funded places;
- the views of the students on the courses offered by the institution—one way of finding this out is through the results of the Postgraduate Taught Experience Survey (PTES); this is rather like the NSS (National Student Survey) for postgraduate students, though, at the time of writing, the completion rates are much lower;
- the place—is it somewhere where you would like to live?

If you are looking for a PhD then this is a much more individualised process, which is likely to be focused specifically on the area of research that fascinates you. Opportunities are often advertised online: e.g. through findaPhD.com or through scientific journals, but often they are obtained through personal contact. During your studies, you will have come across areas of research that have fired your interest, and maybe read papers by researchers in the field. At this stage, then, one of the best approaches is to contact those researchers, either in your own university or elsewhere and ask them if they have places for a research student and if there is likely to be any funding. Reputation is important in research and you should look at the websites for the research groups and see what they are publishing and whether they have ongoing funding from research bodies such as the Research Councils (e.g. MRC, BBSRC, NERC) or from major research charities such as the Wellcome Trust, Cancer Research UK, or Action Research. Talk to the academic staff who are teaching the area you are interested in and see if they can point you in the right direction.

experience, internships, part-time work, or full-time graduate job. Again, this is something that your university careers service can help you with but here's a brief summary.

Advertised vacancies

- employers' websites;
- graduate recruitment websites;
- publications, e.g. The Times Top 100, Guardian UK 300;
- newspapers—national and local;
- scientific magazines;
- professional associations;
- recruitment agencies;
- company profiles on social media, e.g. LinkedIn.

Unadvertised vacancies

- networking—via careers fairs, employer campus visits, and social media;
- personal and professional contacts;
- speculative applications (particularly appropriate for smaller companies);
- business listings and directories (local firms), e.g. yell.com;
- work experience, internships, placements (from which you could hear about vacancies);
- part-time and temporary work (from which you could hear about vacancies).

14.5 Sell yourself

'Selling yourself' is about articulating your suitability and interest for a role throughout the selection process. Being able to 'sell yourself' is a vital aspect of making yourself employable. You have got a good degree, gained experience, and developed your skills, but if you can't articulate your suitability and interest for the role and the organisation you are applying to during the selection process (both formally and informally) then you are unlikely to get very far.

Different employers run their selection processes differently but there are some broad similarities. Smaller employers typically use a two-stage process involving an application or CV plus covering letter followed by an interview or interviews. These vacancies are often less visible (possibly unadvertised), and recruited as needed. Larger employers have much more structured processes and their vacancies are usually highly visible to compete for the best talent. Larger employers deal with a larger volume of applicants and so will use a multi-stage process to screen applicants. Typically the stages larger employers use are as follows:

- online registration;
- online application;

- testing (e.g. situational judgement, logical reasoning, numeracy);
- phone or video interview (sometimes more than one);
- assessment centre;
- interview (often on the same day as the assessment centre and sometimes more than one).

We said earlier that the larger graduate employers may receive hundreds of applications for each position. This highly competitive aspect of the process can leave you feeling somewhat daunted. It's important, however, to be aware of the context: as the graduate recruitment market becomes more competitive, many applicants will reason that they can increase their chances of success by making more applications. This is not necessarily the case. Typically, as the number of applications increases, the proportion of quality applications decreases; this is because the more applications people make, the less time they are likely to spend tailoring them to the needs of the employer they are applying to. It's much better to make fewer, more focused, better-tailored applications to a small number of targeted employers than to churn out lots of generic, only slightly tweaked applications to a large number of employers to whom you struggle to communicate why you want to work for them in particular.

There are many good books and websites available that give information and advice on CVs, applications, and interviews, and your university's careers service will also be able to help you with this (a recurring theme throughout this chapter). What follows, therefore, is simply a summary of the main points to give you an overview.

14.5.1 CVs

It's much more common for applications for graduate jobs to be via application form rather than CV. However, a good, up-to-date CV is a really useful document, not least because you can use it to create a professional online profile. We mentioned LinkedIn in section 14.4.2, Use your network, but to create an online network you need to first have a professional online profile. You can use the information from your CV to populate your online profile (or vice versa if you've already got a LinkedIn profile but not a CV). Even if you don't use a CV to apply for graduate jobs, you will probably use it to apply for part-time or temporary work. Importantly it will also:

- help you reflect on your strengths and areas for development;
- form the basis for your applications;
- act as a useful aide memoire before interviews or assessment centres to remind you of your qualifications, experience, and skills.

These are just the benefits of a CV from your perspective, but what about the audience for whom you are writing: what are employers looking for? A key fact that you need to bear in mind when writing a CV is that employers or recruiters who accept CVs get a lot of them and so they can't spend very long reading them (probably less than a minute). A good CV, therefore, needs to communicate relevant information clearly and succinctly in terms of both structure and content.

Structure

The well-structured CV will help the reader see relevant information quickly and easily to make a quick judgement as to whether they want to read it in more detail. A badly written CV won't even get past this first look. The information needs to be laid out in a clear and accessible way that helps the reader see the information they are looking for. Typically a graduate level CV would include the following headings:

- personal details (name, email, phone number, etc.);
- a brief summary/profile statement (to encourage the reader to read more—highlight your key strengths and make sure you back them up in the rest of the CV);
- education (details of your education to date to an appropriate level, including fuller details about courses relevant to the job—GCSEs would normally be abbreviated to the number and grades, e.g. six 8s (including Maths and English) and three 7s);
- work experience (put your most recent and/or relevant experience first);
- skills (you can put these as part of the work experience or include them under their own heading—make sure you know what kind of skills the employer is looking for and then evidence them accordingly, for example communication, teamwork, digital skills, etc.);
- interests (specify your level of involvement and, if possible, use examples that have some relevance to the job);
- references can just be 'available on request' (usually one academic and one from a work situation—make sure you ask their permission first and share the job description with them so they know what it is you're applying for).

Content

The content of your CV must be tailored to the role and organisation you are applying to. You will have baseline content that you can use as a starting point, but it is vital to adapt the content to the role; think carefully about your experience and skills in the context of what is being asked for. Different people have different tendencies; some will tend to oversell themselves and some will tend to undersell themselves. If you are inclined to oversell, then double-check that you have strong evidence to back up your claims (and certainly don't make things up, this almost always causes problems later); if you are inclined to undersell then now is not the time for modesty! Regardless of where you sit on this spectrum everyone will benefit from a second opinion, so get comments from people who know you well (family and friends) and people who can offer expertise (a contact who knows the organisation or job, or your university careers service). If you find yourself really struggling, it might be simply because you don't have the qualifications, skills, or experience that are required; this is why it's important to start early (see section 14.2), so that you use the time available to you during your degree to make yourself as eligible as possible when you graduate.

> ### Try this—Applying for a graduate position
>
> Find an advert for a graduate job that interests you (your careers service can help you with where to look). Analyse the skills and experience that you think are required for the position and produce an appropriate CV, then get some feedback from your careers service on it.

14.5.2 Covering letters

The purpose of the covering letter is to personalise your approach to an employer so demonstrating that you have taken time to research the job vacancy and the organisation and put some real thought into your application. Your covering letter also provides you with another opportunity to establish your suitability for the job and to emphasise your key 'selling points'. Typically, a covering letter would include the following sections or paragraphs:

- where you saw the role advertised (or if you're not responding to an advert, why you're making a speculative approach;
- what it is that interests you about the role and the company and how you have come to this conclusion (this is where to mention any contact you have had with the company, however brief);
- why you are suitable for the role and the organisation (highlight your key selling points—your qualifications, skills, and experience—and match these against the person specification for the job you are applying for);
- a closing sentence about your availability and that you are looking forward to hearing from them.

In all of these, you will need to convey a genuine interest in, and enthusiasm for, the role.

> ### Try this—Preparing a covering letter
>
> Prepare a covering letter to accompany the CV you created for the previous 'try this' activity.

14.5.3 Application forms

As mentioned earlier, a good, up-to-date CV will form the starting point for filling in an application form. Qualifications and work experience can probably just be lifted from your CV and dropped into the relevant sections of an application form; however, the rest of the form will take more thought and effort. As with CVs, employers will receive hundreds, possibly thousands of applications, so first impressions really count. Make sure that you have researched the role and company thoroughly and that you fulfil the minimum criteria they are asking for. After basic information like personal details, qualifications, and experience, the application form will usually consist of a series of focused questions designed to ascertain specific information. Some questions will require you to provide credible evidence to back up your claims of specific skills and strengths (try using the STAR model to structure your answers: Situation, Task, Action, Result—see Box 14.4); other questions will attempt to establish your

BOX 14.4 The STAR technique

The STAR technique provides a framework to help you answer competency-based questions on application forms or in interviews. Competency-based questions are questions that are usually focused on a particular skill and you are asked to provide evidence that you have performed it well in the past (albeit, probably in a different setting). Examples include:

- Tell us about a time you had to adjust your communication approach to suit a particular audience.
- Give an example of a time when you had to make a difficult decision.
- Describe a situation in which you were working as part of a team. What contribution did you make?

The framework is:

- Situation—briefly set the scene (where and when?)
- Task—what was the aim and what was your role?
- Action—what did YOU do and how did YOU contribute (this is the main part of the answer)?
- Result—what results were achieved (you will want to be positive here but it's also good to say how the result could have been improved)?

Example:
Question: Tell us about a time you had to adjust your communication approach to suit a particular audience.

Answer:
- (Situation) I'm a student ambassador for my university and that means that I occasionally speak at local schools about what it's like to study at university.
- (Task) I was asked to talk to a group of year 10 students about what it's like to study at university.
- (Action) I was given the slides that had been used previously but when I looked at them I decided that I wanted to do things a bit differently. I knew I was only going to be talking to a group of about five students, so a PowerPoint presentation seemed a bit over the top. I wanted to make it less formal and encourage more interaction. So, I decided to run a much less formal question and answer session as I thought they would probably have questions from the larger group presentation that they would be attending before mine.
- (Result) And I'm pleased to say it worked really well! Four of the five students asked questions and I seemed to strike up quite a rapport. Perhaps next time I might ask them to write a question each to encourage the quieter ones a bit, but other than that I'd probably do the same again.

motivations for applying and your suitability for the role and organisation. Just like an essay, a good application takes a long time to research and write, so make sure you give yourself plenty of time. And as with CVs, second opinions are really helpful, both from people who can comment on your skills and people who can advise you on the role or organisation, so make sure you seek out help.

14.5.4 Testing

As part of the selection process, many employers use aptitude tests and personality questionnaires to assess suitability. Such tests are structured ways of evaluating how people perform on various tasks or react to different circumstances.

Aptitude tests aim to measure your competencies and intellectual capabilities as well as your logical and analytical reasoning abilities in a particular area. They aim to assess your abilities to use specific job-related skills and to predict subsequent job performance. The most commonly used tests are:

- verbal reasoning;
- numerical reasoning;
- situational judgement.

Aptitude tests, as with any part of the selection process, can be quite nerve-wracking, but perhaps particularly so because of their unfamiliarity. The best way to increase both your confidence and performance is to practise. Ask your university careers service for help and for links to online practice tests.

Personality questionnaires explore the way in which you do things and how you behave in certain circumstances. They can assess your preferences, motivations, interests, values, and attitudes. You should give careful thought to your responses but, for personality questionnaires, there are no right or wrong answers (hence it being called a 'questionnaire' rather than a 'test'). It may be tempting to try to respond to personality questionnaires in ways that you think will attract the prospective employer, rather than necessarily reflecting your actual personality traits. This is not such a good idea as the questionnaire may (certainly the more sophisticated ones will) include questions that evaluate internal consistency across the questionnaire as a whole and this is likely to pick up on any areas where you are creating an unrealistic portrayal of yourself.

For both aptitude tests and personality questionnaires, make sure that you give yourself time and space and a quiet, distraction-free environment to complete them in one sitting.

14.5.5 First-stage phone or video interview

As we mentioned earlier, larger employers use multi-stage selection processes to deal with large numbers of applicants; this often includes either a phone or video interview, or more commonly now, a recorded video interview after the testing but before the assessment centre or in-person interview stages.

First-stage phone or video interview (synchronous)

A phone or video interview is usually pre-arranged so you know when to expect it, but can be unannounced; either way, some preparation is essential, just as it is for an in-person interview. In this section, we are referring to a *synchronous* call, that is to say the candidate and the assessor are on the call at the same time.

Assuming the call is prearranged, you can prepare your environment accordingly to ensure that you have access to relevant information and that the space in which you take the call is quiet and free from distractions. Make sure you have a copy of your application form and CV to hand (even if you haven't submitted a CV, it's a useful summary of key points) and let the people you live with know that you are expecting a call and so don't want to be disturbed. If you get an unannounced call and you don't have relevant documents to hand or the environment is too noisy or distracting then ask if you can arrange a more suitable time to take the call—you are not obliged to speak there and then if the call was unexpected.

In addition to preparing your environment you should also prepare for the interview itself. Preparing for telephone or video interviews is very similar, however, to preparing for face-to-face interviews, so see section 14.5.7 *Interviews* for more information on this.

During the interview itself you will need to be alert and well prepared. If it's a phone interview the biggest and most obvious difference is that there are no visual clues for either you (the candidate) or them (the assessor); this means that the tone and rhythm of your voice become much more important. If you speak in a monotone voice, this will be more of a problem on the phone as there will be no facial expressions to help the interviewer interpret the meaning of, or the emotions behind, what you are saying. The following tips will help.

- Smile! Silly as it might sound, smiling whilst you talk really helps. You will come across as more friendly and confident. Use gestures as in normal conversation and be enthusiastic where appropriate.
- Be organised—have somewhere to take notes and your calendar to hand in case they like the sound of you and want to invite you to a meeting.
- Try not to be put off by pauses from the interviewer—they may be taking notes (this happens in in-person interviews too).

Remember a phone interview is a precursor to an in-person interview—it is important to create a good impression and answer the questions as best you can. Don't fall into the trap of under-preparing for a phone interview because it seems less formal or less important than an in-person interview—it is an important stage of the selection process and can be the point at which you are excluded from the process or the point at which you are flagged up as a strong potential candidate.

Almost everything described above for phone interviews also applies for an interview conducted as a video call via, for example, Teams or Zoom. Additionally, with a video call, you will also need to think about your appearance and your backdrop, see the next section for more information on this.

First-stage recorded video interview (asynchronous)

It is now increasingly common for the larger graduate recruitment schemes to use *asynchronous* video interviews; this is where the interviewee is recorded responding to pre-set questions, and it is not conducted in real time. Candidates are invited to an 'interview' via a web link and are then invited to respond to pre-set questions, which will be reviewed at a later stage by an assessor. Employers are increasingly using this format instead of phone interviews as it provides the added visual element, can be easily reviewed, and (importantly) doesn't require the assessor to be available at the same time as the candidate.

Once you are signed in to the video interview platform the format is typically:

- you are asked a question;
- you are given a minute or so to prepare your answer;
- the software then records you responding to the question;
- you then move on to the next question.

The precise details will depend on how the system has been configured, so practice is important. Your university careers service will be able to provide help, which often will include an opportunity to practise video-based interviews (and play them back to yourself so you can assess your own performance) and even provide a bookable room in which to have your interview.

Much of the advice detailed above for phone interviews applies, but in addition you should also consider the following:

- Prepare your environment—clearly a video interview allows you to be seen as well as heard, so think about how the 'set' looks (Is it well lit? Is it tidy? Is it quiet?). Check the technology is working beforehand, and if you are unsure about using your own room then try to book a room in you careers service.

- Prepare for the interview—in addition to preparing for likely questions (see 14.5.7 *Interviews*), you should also dress appropriately. What 'appropriate' means will depend on the context, but it's always better to be slightly too smart than not smart enough (you won't need to polish your shoes but dressing the part will probably help).

- During the interview—the software will probably allow you to practise your answers and play them back so you can review them, so make sure you take account of this. Use body language appropriately, and don't forget to smile!

14.5.6 Assessment centres

Many employers use extended selection exercises as part of their recruitment process; these are often run as an assessment centre. Assessment centres are one of the most reliable indicators of future job performance and usually involve a series of exercises designed to show the selectors whether or not you possess the skills required. Typically assessment centres include the following elements (assuming testing—section 14.5.4—has already been part of the selection process):

- company presentation;
- group tasks;
- individual tasks;
- interview.

The company presentation is usually at the beginning of the day, immediately following an introduction and welcome. You should have researched the company well already as part of the previous stages of the selection process, but it is important to remind yourself of what you have already discovered. During the presentation you need to pay close attention to what is being said and perhaps think of a question that you would like to ask at the end, should you be invited to do so. The key points made during the presentation will probably be useful to you for other aspects of the day so make sure you are concentrating. You may find it helpful to take brief notes—but don't let taking notes distract you from what is being said.

The type of group tasks used will vary considerably, depending on what the assessors are trying to assess, but it will usually be a discussion, a role play, or a business game. In a discussion, groups are given a topic to discuss and will be expected to agree a response and present back their conclusions. Assessors will observe candidates' interpersonal skills and the contribution everyone makes to the group. In a role play, you could be asked to play a role in a simulated exercise. The assessor will look at how you put your views across, whether you can construct a convincing argument and how well you influence and persuade others. A business game will involve tackling a simulated business situation where a task has to be performed, targets met, and difficulties overcome.

These exercises often feel a bit artificial and can make you feel quite self-conscious, but it is important to try to involve yourself as fully as possible and make a positive contribution to the group. It's not about talking the most or being dominant, neither is it about hanging back and expecting everyone else to do the work; it's about putting forward your ideas in a convincing and constructive fashion whilst encouraging other people to do the same.

Individual tasks might take the form of an inbox exercise that might involve reading and prioritising information, identifying action required and any deadlines, drafting replies, preparing reports, and other actions. Alternatively you might be asked to give a presentation: you may be given the topic in advance (so you can prepare) or you might be asked to prepare it on the day. Regardless of whether you have advanced warning or not, the key to success lies in anticipating your audience's needs, delivering within the specified time frame, and communicating clearly and concisely.

Finally, you will need to be prepared for an interview as part of any assessment centre. Even if you had a phone or first stage interview previously you need to expect more in-depth questioning. Your knowledge, your ability to do the job, and your motivation will be under close scrutiny. For information on preparing for interviews see the next section.

Throughout the assessment centre, you should assume that every aspect is assessed. You will need to arrive in good time, be polite to everyone, and join in, even over lunch. If you find small talk difficult ask other candidates about their courses or the assessors about how long they have been with the company. When presented with new material for exercises, take time to make sure you understand the key information and the requirements of the task. Try to stay calm and focused throughout; don't dwell on any mistakes, just concentrate on doing well in the next task. After the assessment centre, make notes on your experience and ask for feedback on your performance.

14.5.7 Interviews

Interviews are usually the final stage of the selection process and the culmination of all your hard work up to this point. Without giving a good interview, you won't get the job, but to get an interview you will have needed to have shown suitability to, and potential for, the role. Interviews are always nerve-wracking experiences, but you can use the nerves to your advantage to help you stay sharp and focused. You also need to stay positive—the organisation must have some confidence in you for you to have got to this stage and you need to continue to give them reasons for that confidence.

There are many books and websites giving advice about how to perform well at interviews. The main points though are to do your research, to practise answers to possible questions, and to prepare some questions to ask.

You will have had to have done some research to get to the interview stage of any selection process and it is important that this information is fresh in your mind at interview.

It can be easy to think you have done your research already because you did that at the beginning of the selection process but you will need to do more prior to the interview itself. The company's website is the place to start when researching a company, but stories about the company in the news (try a web search for the company and then choose the 'news' results) plus information you can glean from meeting in-person with them (see section 14.4.1) are also useful sources of information.

> ### Try this—Researching a company
>
> Choose a company that you think you might be interested in working for—it could be big or small, well known or not. Do some background research on the company that includes: what they do, how many people they employ, where they are based, whether they have overseas sites, what their newest product or service is.

Most of the interview will be taken up with answering questions and you need to allocate your preparation time accordingly. You will need to read and re-read your application and CV because you will doubtless get asked about some of what you have written at the early stages of the selection process. You also need to have practised answers to questions such as:

- Why have you applied for this role?
- What makes you suitable for this role?
- What are your strengths?
- What are your weaknesses?
- Describe a difficult situation and what you did about it.
- What is your proudest achievement?

You will need not just to think about how you will respond to these and other questions but also practise your responses out loud. Ideally, practise with someone who can give you feedback on how you are coming across and ask you some follow-up questions—a friend or parent or careers adviser.

The final stage of the interview is usually concerned with questions you have for them. It is important not to clam up at this point; try to think of a couple of questions before the interview. Some of your questions may well be answered during the interview, in which case you can say that these have been answered, but it is good to be able to have one or two questions to ask. Ask questions that show you have thought about the job and the company.

Sometimes final-stage interviews are conducted remotely via a video call. This isn't common for graduate recruitment other than earlier in the selection process (as described in 14.5.5 *First-stage phone or video interview*, where specialist video interviewing software is increasingly used), and, if you're given a choice, we recommend that you attend in person (they should reimburse reasonable travel expenses incurred). However, if you are interviewed via a video call you will need to consider the following:

- make sure the technology is set up and working well in advance so you have one less thing to worry about immediately before the interview;

- ensure that you won't be disturbed, again your university careers service may allow you to book a room for the purpose;
- consider the 'set' for the interview—well lit, tidy, and professional is probably best;
- dress professionally—whilst this may not seem as important is in an in-person setting, dressing the part will help you be more mentally prepared.

In both your answering and asking questions, in fact throughout the whole selection process, you need to present yourself well, be friendly, and be positive. If it doesn't work out and you are unsuccessful at this or any stage of the selection process, ask for feedback. There will always be room for improvement and, given how much time and effort is involved in completing a selection process and how much is riding on it, it is important that you use feedback to improve your approach and increase your chances of success. It's very unlikely that you'll be successful in the first job you apply for, so try and look on interviews as good practice, rather than success or failure; it's worth applying for jobs and attending interviews for the experience you gain to reflect on and improve your approach next time.

 ## Chapter summary

Making yourself employable is a vital part of your degree. It is essential that you start thinking about your employability from the beginning of your degree, rather than at the end, so that you can enhance your employability as you go along rather than trying to cram it all in at the end. Whilst getting a good degree is an important part of finding a good job, it represents only a fraction of what employers look for in potential candidates. Since employers can easily find out about your academic qualifications (they just ask you to tell them what you've got), the selection process is weighted towards finding out the skills, experience, and attributes you can offer an employer in addition to your academic qualifications. Gaining experience, making contact with employers, and selling yourself are essential elements of making yourself more employable.

 ## Reference

The Future of Jobs Report 2023. (2023). Retrieved from https://www.weforum.org/publications/the-future-of-jobs-report-2023/

Index

Page numbers in **bold** indicate the start of a chapter on the subject. Tables, figures, and boxes are indicated by an italic *t, f,* and *b* following the page number.